Urban Revitalization

Following decades of neglect and decline, many US cities have undergone dramatic regeneration. From New York to Nashville and from Pittsburgh to Portland, governments have implemented innovative redevelopment strategies that respond to a globally integrated economy and cope with declining industries, tax bases, and populations. However, despite the new amenities in revitalized neighborhoods, spectacular architectural icons, and pedestrian-friendly entertainment districts, the urban comeback has been highly uneven. Thriving cities are defined by a divided population of highly educated, well-paid professionals and a low-wage workforce while many cities and neighborhoods continue to seek pathways to improve their conditions. It is crucial that students of urban revitalization recognize contemporary challenges, possible courses of action, and their effects on diverse people and places.

Urban Revitalization is the first textbook to comprehensively and critically synthesize the approaches and challenges involved in urban revitalization. The book is divided into five sections. The introductory section defines urban revitalization and explores the disparities across the US. Section II examines the historical causes for decline, discussing the factors that shape contemporary revitalization initiatives and outcomes. Section III offers a critical examination of contemporary urban revitalization policies, strategies, and projects. This section also provides a rich set of case studies that contextualize key themes and strategic areas across diverse contexts including the urban core, central city neighborhoods, suburban areas, and shrinking cities. Section IV introduces analytical techniques and key data sources for revitalization planning. The book concludes by evaluating the current state of revitalization planning and reflecting on the emerging challenges.

Urban Revitalization integrates academic and policy research with professional knowledge and techniques, combining a critical examination of best practice and innovative approaches with an overview of the methods used in urban revitalization. Thoughtfully organized to provide graduate and upper-level undergraduate students with a comprehensive and accessible resource, this volume will also serve as an essential reference guide for regeneration professionals.

Carl Grodach is a Senior Lecturer in Urban and Regional Planning at Queensland University of Technology, Australia.

Renia Ehrenfeucht is Professor of Community and Regional Planning at the University of New Mexico, USA.

URBAN REVITALIZATION

Remaking cities in a changing world

CARL GRODACH AND RENIA EHRENFEUCHT

Routledge
Taylor & Francis Group

LONDON AND NEW YORK

First published 2016
by Routledge
2 Park Square, Milton Park, Abingdon, Oxon OX14 4RN

and by Routledge
711 Third Avenue, New York, NY 10017

Routledge is an imprint of the Taylor & Francis Group, an informa business

British Library Cataloguing in Publication Data
A catalogue record for this book is available from the British Library

Library of Congress Cataloging in Publication Data
Grodach, Carl, author.
Urban revitalization : remaking cities in a changing world / Carl Grodach and Renia Ehrenfeucht.
Includes bibliographical references and index.
1. Urban renewal--United States. 2. City planning--United States. I. Ehrenfeucht, Renia, author. II. Title.
HT175.G756 2016
307.3'4160973--dc23
2015027514

ISBN: 978-0-415-73053-2 (hbk)
ISBN: 978-0-415-73054-9 (pbk)
ISBN: 978-1-315-85021-4 (ebk)

Typeset in Times and Bell Gothic
by Saxon Graphics Ltd, Derby

Printed and bound by CPI Group (UK) Ltd, Croydon, CRO 4YY

Contents

Figures

Tables

Boxes

SECTION I

Changing Regions, Local Lives
Urban Revitalization in Context

1

Introduction
Definitions, concepts, and geographies

Following decades of neglect and decline, many US cities have undergone a dramatic renaissance. From New York to Nashville and from Pittsburgh to Portland, many cities have engaged in innovative redevelopment strategies to adapt to a globally integrated economy and cope with transforming industries and shifting demographics. Along the way, new forms of governance and redevelopment partnerships have emerged to realign traditional power structures dominated by downtown property interests and to give voice to underrepresented communities.

However, despite many gains, the urban comeback has been highly uneven. Thriving cities are often defined by a bifurcated population of highly educated, well-paid professionals and a low-wage workforce. Diverse and vibrant immigrant neighborhoods have developed alongside those that are economically and socially segregated. While cities have transformed high-profile brownfield sites, disadvantaged neighborhoods continue to grapple with abandoned and environmentally contaminated land. Moreover, urban cores in growing cities are flourishing, gaining spectacular architectural icons, pedestrian-friendly entertainment districts, and new housing. At the same time, surrounding inner-ring suburban areas face mounting problems, and shrinking cities wrestle with long-term decline.

At the end of the 2000s, the great recession brought this into sharp relief. While some communities barely felt the recession, others experienced inordinately high rates of foreclosure and subsequently left residents to deal with growing residential and commercial vacancy, property abandonment, and anxiety about declining neighborhood conditions. This situation was particularly intense in communities of color where historically residents have been underserved by banks, have poor access to credit, and fewer opportunities to become financially literate. These neighborhoods became a target for predatory lenders. As a result, African American borrowers were nearly three times more likely, and Latino borrowers twice as likely, to receive a subprime mortgage as white borrowers from a mortgage lender that went bankrupt during the financial crisis (Avery, Brevoort, and Canner 2007).

The conditions are further intensified in shrinking cities already contending with the legacy of industrial decline. In neighborhoods like Cleveland's Slavic Village, vacancy rates climbed from 13% to 33% and rates of poverty grew from 27% to 43% in just six years (Ferenchik 2015) (Figure 1.1). Clearly, these neighborhoods face difficult challenges. In order to address them, we need to understand the array of factors that bring them about and the possible routes toward revitalization.

Figure 1.1 Abandoned property in the Slavic Village neighborhood, Cleveland, OH
Photograph by Gus Chan, The Plain Dealer

Urban Revitalization: Remaking Cities in a Changing World examines the processes that shape these patterns of disinvestment. We study the many efforts of the people and organizations that work to strengthen and revitalize their communities. The book tackles complex questions: Why do unequal outcomes persist along lines of race and income? Why do major global events like the financial crisis of 2007 impact communities differently? Can local communities initiate urban revitalization in the face of seemingly uncontrollable global forces? Why do major investments go to some places, while others are bypassed? How can urban policy and planning efforts lead to better outcomes?

Urban Revitalization explores these questions and challenges across multiple contexts. We study urban revitalization in central city neighborhoods, older industrial areas, emerging immigrant neighborhoods, and inner-ring suburbs. We examine revitalization efforts in large, established metropolitan areas, sprawling Sunbelt cities, shrinking cities, booming housing markets, and those in decline. The aim is to provide a comprehensive basis for untangling the complex and divergent impacts of revitalization efforts in these diverse contexts and make them comprehensible to readers becoming familiar with urban revitalization processes. Throughout, the book provides a critical examination of established practices while exploring innovative and emerging approaches to revitalization. We ground this knowledge in the history of urban redevelopment and the broader forces that impact policy and project outcomes. In addition, the book introduces readers to the methods and data used to document and understand local situations and urban revitalization processes.

In the following sections, we begin by exploring the meanings of urban revitalization and the factors driving urban change to help illuminate the key issues and conflicts surrounding revitalization efforts.

DEFINING AND PROBLEMATIZING URBAN REVITALIZATION

In the most basic sense, the term revitalization refers to a rebirth or revival in the conditions and character of a place that has endured a period of decline. While cities and neighborhoods may seem to change little on a day-to-day basis, in reality they are in a state of constant flux.

Movements of people and jobs, new construction and demolition, changes in land use, zoning and infrastructure investment mean that the opportunities available in cities and neighborhoods vary over time.

For myriad reasons, which we explore below and in the following chapters, neighborhoods or even entire cities and regions may undergo a process of urban decline. Residents and businesses begin to move out of the area, unemployment mounts, and housing and infrastructure maintenance is deferred as municipal revenues and household incomes decrease. As real estate values and the population decline, school and municipal revenues decrease, public schools struggle, and urban services worsen. As retail establishments close and municipal employment decreases, more jobs are lost. This chain of problems poses serious challenges for the remaining residents. Public officials, business and community leaders, and ordinary residents work to improve these conditions and, partly due to their efforts, cities and neighborhoods can make a comeback. Both decline and renewal are impacted by a confluence of factors including government intervention, changing real estate and industry dynamics, private sector investment, and community efforts.

As we will elaborate, this is not a cycle isolated to specific places. For example, when US cities began to lose their traditional manufacturing employment in the 1970s and 1980s, the vast majority of these jobs went to lower-wage countries. Many regions experienced economic and fiscal stress, which was clearly visible in the deteriorating residential and commercial districts that continue to define many parts of the urban core and inner suburbs. In some places, however, these conditions have been offset as immigrant entrepreneurs and families established new businesses and communities. Their personal and financial investments have begun to turn around challenged neighborhoods, as we show in chapter 8. In other words, decline and renewal is a local phenomenon determined in part by the action of people in a specific place, but it is also tied to global forces such as deindustrialization and immigration.

Just as people move and places change over time, the idea of urban revitalization has taken on different meanings depending on the time, place, and agenda. As we discuss in chapter 3, federal revitalization policy came to prominence in the 1950s under an urban renewal policy concerned with large-scale, wholesale demolition and reconstruction of entire central city districts. On and off again since the 1960s, this approach has been tempered with attempts to incorporate community participation into the planning process but, at the same time, it has been marked by an increasing focus on market-based outcomes and initiatives over which residents have less control.

The term urban revitalization took on a different tone in the 1980s as cities began to respond to global economic changes and fiscal austerity at the federal level (see chapter 4). During this time, urban programs explicitly focused on building places to attract upwardly mobile professionals as a means of establishing or maintaining their economic competitiveness, rather than as a means of creating broad-based opportunity for groups marginalized by larger economic and political forces. In the 1990s, this stance shifted somewhat as cities and states began to blend a social agenda into their market-based programs. For example, we see a rising emphasis in the US on building mixed-income housing to turn around struggling central city neighborhoods (see chapter 7); and the UK's "Third Way" programs that attempted to arrest social exclusion through community participation initiatives (see Tallon 2013 on urban revitalization from a UK perspective).

Throughout these transformations, the focus of revitalization policy has largely been on *place based* initiatives intended to alter the built environment and catalyze positive social and economic change. Place-based strategies focus on improving the conditions in specific neighborhoods or urban areas as a route toward improving people's lives. They differ from *people-based* strategies that attempt to directly assist individuals and families irrespective of location. People-based programs include housing vouchers that help people rent housing in different areas or "food stamps," the Supplemental Nutrition Assistance Program among others. Place-based approaches to urban revitalization can and should take account of these issues as well by considering how to improve conditions for people living in an area. Even cities and neighborhoods suffering high levels of disinvestment continue to have people who live and work there, and local conditions greatly impact their opportunities. Because of this, place-based urban revitalization is a critical dimension for urban residents' well-being.

While large-scale place-based improvement projects continue to dominate the urban revitalization agenda, following the fiscal crisis of the 2000s the scale of change has often downsized to emphasize smaller-scale projects. Simultaneously, while revitalization programs continue to target redevelopment of the built environment, calls for social inclusion and healthy, sustainable environments are increasingly part of the revitalization discourse.

The purpose of urban revitalization programs is to build places that enable diverse people to thrive and prosper. Critically observing past revitalization initiatives—both the successes and adverse outcomes—shows us that robust urban revitalization has six dimensions:

- **human capital:** building the skills and capacity of individuals to take advantage of educational and employment opportunities
- **social–cultural equity:** dismantling racism and other forms of structural inequality as well as fostering community interaction and collaboration, cultural identity, and representation
- **built environment:** developing quality, affordable housing for diverse incomes, basic and specialized infrastructure, and healthy environments
- **place attractiveness:** improving urban design, public spaces, and consumer services
- **economic competitiveness:** attracting, developing, and supporting businesses, workforce development, and entrepreneurs
- **environmental sustainability:** reducing adverse environmental impacts and restoring ecological systems that support people and other living beings.

Today, urban revitalization has become something of a buzzword. It is a broad and malleable term imbued with politically charged meanings that media, government, community groups, and academics alike employ in different ways for different ends. A Google image search of "urban revitalization" suggests that revitalization is defined by new, relatively dense, and often sterile shopping and residential areas where pedestrians dominate (Figure 1.2).

In reality, revitalization outcomes are considerably more complex. While cities may tout flashy new buildings and pedestrian-filled streets as signs of urban renewal, there is much more to the promotional images and idealized visions of urbanism. When we look at these images, we do not really know how they came to be, or if the residents and businesses there prior to redevelopment are still around. Place-based improvements are certainly important, but they must occur in tandem with initiatives that assist the people who live in the area and ensure they can enjoy and benefit from the positive changes that occur. Moreover, many neighborhood improvements may not be immediately visible or photogenic. A community center may provide a new day care or job readiness program, or a development corporation may build modest but quality affordable housing where it is lacking. These improvements can contribute to improved living conditions and enhanced employment opportunities for residents. They likely will not make the headlines like an abandoned elevated rail line transformed into an urban park, but they can make a difference in people's lives.

PINPOINTING THE FACTORS DRIVING URBAN CHANGE

Understanding where, why, and how urban revitalization occurs requires considering local action in a broad context. Local planning decisions and project outcomes depend on factors that cut across a variety of scales. Both national and global trends affect local planning decisions. Local actions and ideas also travel and are replicated in surrounding communities and in cities and countries around the world. Moreover, revitalization planning engages diverse actors. Planners, mayors, community representatives, residents, developers, and chambers of commerce all wield

Figure 1.2 A rendering of "urban revitalization". Auraria redevelopment project, Designer Mithun
© Dennis Allain (2007)

varied resources and often have different interests. Technical expertise and political manoeuvring can steer development to places in need of revitalization or to those areas with the highest potential gain from public investment. Policy can intervene to rectify market-based inequalities or focus primarily on generating financial returns. In short, the political and economic context for revitalization matters a great deal.

In this book, we examine revitalization and redevelopment from the top down and the bottom up through an integrated focus on the political economy of cities and an analysis of socio-cultural factors. This approach sheds light on the ways in which local economic and political institutions work within regional-, state-, and global-level conditions to shape policy and programmatic agendas. In conjunction, it recognizes the roles of social trends, cultural factors, and grassroots actions in shaping urban revitalization outcomes. In addition, we highlight how structural inequality shapes and impedes effective action to create urban conditions where all people can thrive. Examining revitalization through this integrated lens enables us to comprehensively account for the factors that define policy priorities and produce workable policy strategies, and—ultimately—the types and location of revitalization projects and their impacts on different communities.

UNEVEN DEVELOPMENT

The processes of growth and decline are interrelated. As some places see real estate values increase and gain access to more jobs and services, others struggle to maintain local opportunities. The processes of urban growth and decline are integral to a market economy in which urban space and the built environment are treated as products like any other commodity from cars to shoes (Lefebvre 1991). Many actors in the urban development arena—including property investors, business interests, and the public sector—consider places primarily based on their actual and potential monetary worth or their exchange value. A key determinant of the urban development process, therefore, is the investment in the built environment as real estate speculators, business and industry interests, and government seek to create surplus value or, in other words, to profit from their investments.

Because this is a speculative process, urbanization is characterized by uneven development (Harvey 1985). Speculators invest (or disinvest) in a place because they anticipate a profit from their decisions. We can observe this basic process at work in different types of places and time periods, as we discuss in more detail in chapter 2. Every phase of urbanization, from the industrial city and the early suburbs to downtown urban renewal and the gentrification of central city neighborhoods, is driven in part by a shift in investment among places. Speculators may shift their focus because of new spatial demands and social trends. Over a longer period, changes in the larger economic structure from heavy manufacturing to an economy driven by knowledge and information have reshaped the built environment as it generates new work and living patterns and new development demands. This also has regional effects. In the 1980s and 1990s, major channels of investment shifted resulting in the growth of Silicon Valley and the decline of auto and steel centers such as Detroit and Pittsburgh. The competitive drive to extract profit from the built environment created cycles of decline and renewal as capital seeks new areas for investment, continuously shifting the geography of urban growth both within and between cities.

This process of speculative investment and disinvestment has been called "creative destruction" (Schumpeter 1950, Harvey 1985, 1989). As Harvey (1985: 83) explains, this is "a perpetual struggle in which capital builds a physical landscape appropriate to its own condition at a particular moment in time, only to have to destroy it, usually in the course of a crisis, at a subsequent point in time." The process of creative destruction describes the way in which economic systems are reconfigured as investors attempt to draw profit from urban land. The result is an often slow but constant transformation of cities' built environments and the available options for how and where we live and work. The city then is a reflection of the production and consumption decisions made by people over different time periods.

Nonetheless, the growth and change of cities is not simply a speculative process. While creative destruction helps to explain why some places experience disinvestment yet become future targets of reinvestment and revitalization over long periods of time, people's attachments and the persistence of the built environment complicate the situation. As Logan and Molotch (1987) explain, places have both exchange value *and* use value—the social, communal, and personal attachments people have to place. People's histories, memories, and cultural traditions also play an important role in how places develop and can even influence development. Harlem in New York, the French Quarter in New Orleans, and San Francisco's Castro District are all places where use and exchange value collide. Each of these areas has a very strong identity and place attachment, but they have also experienced their share of speculative development, driven partly by their unique qualities. While the same can be true for the more ordinary places where most of us live, many communities struggle to gain new investments.

The need for revitalization comes about because development is uneven and, often, the drive to extract exchange value from a place comes at the expense of the use value for many residents. Despite disinvestment and decline, people continue to be attached to their cities and neighborhoods. Some attachments are financial—and business interests and outside speculators consider the best ways to profit off the land—but other attachments extend beyond. People establish roots in particular regions, cities, or neighborhoods, and they have personal attachments to both place and people. Picking up and moving to a new home is not an easy decision and many people choose to stay and defend and improve their communities rather than move on. However, the capacity to do so varies. Residents in neighborhoods like Slavic Village, for example, face a host of obstacles that restrict their potential to effect change. Ultimately, urban revitalization requires working within the context of uneven development and balancing different interests to create good living conditions for residents and stabilizing areas with weaker growth potential and access to resources.

MULTI-SCALAR INTERACTION

The processes that lead to uneven development and the outcomes of urban revitalization agendas are shaped by interactions among global economic forces, national policy, and local action. Most revitalization initiatives stem from an array of place-based actors who work locally. They may attempt to position a city or region within the global economy, or seek to create desirable living conditions for ordinary citizens and counter the inequitable and unwanted effects of decline and growth. The local context is nevertheless defined by the intersection of a dense network of economic and cultural factors ranging from globally connected financial institutions and migration networks to state and federal legislative frameworks. Local actors both work in conjunction with national and global actors, and respond to and shape national and global trends.

An integrated global economy has resulted in patterns of industrial location where manufacturing and management are located in different world regions. Corporations have become networked throughout the world, connected through flows of goods and information. Large corporate developers and individual speculators alike engage in global land investment and development in search of profitable sites around the world. Likewise, people migrate to work and seek opportunities and, in doing so, they alter settlement patterns. They bring their culture and customs with them, transforming neighborhood character and creating new businesses. This is the case not only in global cities such as Los Angeles and New York, but increasingly in Rust Belt cities like Detroit and St. Louis and small towns across the Midwest and Southern US.

National policy also influences the external resources and options available to cities and regions, and national governments offer different amounts and types of aid to cities. US cities receive relatively few national resources for urban development and social services when compared with peer countries. Nevertheless, the local context is an influential factor in shaping urban change and revitalization because projects and initiatives are arranged and administered locally.

Within a given country, cities and regions are differently positioned and their situation reflects local advantages and challenges. Localities have different characteristics and resources, from their natural and built environments to the local development climate. Local actors, including city officials, developers, and community groups, decide what projects to put forward and where they will be located. This creates an urban area's character filtered through national and global influences. Local revitalization efforts therefore both shape broader trends and face limitations from global trends and actors.

MULTI-SECTOR ACTORS AND COALITIONS

Working across these scales, people rebuild local environments through the formal and informal institutions they create. Formal institutions include governmental systems and agencies, laws that govern property, development, and trade among the myriad other ways we organize social and political lives. Informal institutions refer to customs, behavior in organizations or communities, and other social expectations that shape what occurs. In other words, development is not singularly dictated by an inexorable market logic, but is a contested, dynamic, and therefore political process. Because local outcomes are shaped by people, institutions, and ideas interacting among these different scales and regions, local actors must focus on what happens in a given neighborhood and respond strategically to global and national trends.

All urban revitalization initiatives come about through the efforts of coalitions that include public, private, nonprofit, and community sector actors. Within each of these sectors, different people and agencies also have varied priorities. Because people and the businesses and organizations to which they belong are not bound to a singular place, local interests involve national and international participants.

The public sector collectively describes many different people. It includes elected public leaders such as the mayor and council members but also has numerous agencies—from economic development to housing and planning—that have an interest in urban revitalization. In the US, state offices also influence local action and much funding comes from federal agencies such as Housing and Urban Development (HUD), the Department of Transportation (DOT) and the Environmental Protection Agency (EPA). The public sector provides services to local residents, regulates activities to protect community health and welfare, but also looks outwards to compete effectively in the global economy.

The private sector is equally diverse. It includes large corporations and local businesses that are interested in local workforce and property. Housing developers might work locally or nationally. Financial institutions contribute to determining what can be built. Property owners and local businesses also have a stake in the communities as well as their profits.

Nonprofit and community-based organizations, also called the third sector, stand outside of government and the private sector. The third sector can include community and neighborhood organizations as well as national philanthropic organizations such as the Local Initiatives Support Corporation (LISC) or Enterprise Community Partners. Third sector organizations can instigate and fund revitalization projects. They can also get involved as community members. In some cases, organizations start in response to a threat, including the threat of urban renewal or other revitalization initiatives. As we discuss in chapters 5 and 14, there are many different ways that community members formally participate in revitalization.

Traditionally, the public and private sectors work together, forming what Stone (1989) has called an urban regime. Local governments depend on building a coalition with other sectors to establish their capacity to govern (Stone 1989, 2005). Logan and Molotch (1987) have argued that these actors work together to promote urban growth, working as a growth machine (see also Molotch 1976). They describe the growth machine as a coalition of interests rooted in a particular place and united under the banner of local economic growth. In this view, government, major property owners and developers, financial institutions, and media all work together to determine patterns of growth and development because they all benefit. These theories help to explain the formation of public–private partnerships that enable private sector interests to steer revitalization effort to some areas and away from others. They help us understand both the ambitious postwar redevelopment projects we discuss in chapter 3 and the new alliances in the era of privatism and neoliberal governance discussed in chapter 4.

However, as we elaborate in chapters 4 and 5, contemporary federal governance and a globally connected neoliberalized economy alter cities' opportunities. While we certainly continue to see partnerships between local governments, real estate, and industry to orchestrate complex revitalization schemes, we also see revitalization programs led by a much wider range of actors including universities, hospitals, cultural institutions, and philanthropic foundations, often in conjunction with community interests. With a broader set of actors also come more complex agendas that may encompass human capital, cultural, and quality-of-life issues alongside property development (Clark 2004).

As a result, revitalization initiatives cut across multiple governmental agencies—community development, housing, economic development, health and sanitation, and even cultural affairs—and incorporate partnerships with an array of private and community sector interests. Revitalization efforts commonly have strong public sector involvement, but programs also occur through informal place- and community-based interests without official government endorsement.

SOCIO-CULTURAL TRENDS AND URBAN INEQUALITY

Urban development is also shaped by social and demographic trends. Migrations, or moving to or from an area, are a primary factor that shapes how localities change. Decisions to stay or move at the individual and household level can be influenced by many factors including job opportunities, attachments to people and places, environmental change, or change in neighborhood conditions. Intraregional moves, such as between central city and suburbs, have profound effects on the communities within them. Interregional and international migrations both reflect and influence a city or region's opportunities and prospects. It is important to recognize that what appear to be local transformations often come about through uneven development occurring far away as people enduring economic hardship, for example, choose to leave their homes. As we discuss in chapter 8 in the context of suburban revitalization, the movement of people between places also creates new possibilities and tensions. These moves establish entirely new and sometimes unseen networks of labor, finance, and culture among formerly unrelated places.

Social trends refer to changing patterns of living or working. Social trends may occur in response to economic and environmental changes as well as policy initiatives, and the trends themselves can function as the basis for change. For example, the expansion of the internet has had a dramatic impact on how people communicate and how work gets done. Climate change is causing us to rethink how we build our cities. Changing demographics such as the growth of aging baby-boomer and immigrant populations also affects urban labor and housing markets. Rising income inequality contributes to neighborhood segregation and, by extension, equal opportunity. Initiatives to instigate urban change will be impacted by social trends and, as a result, it is important to pay attention to changing migration and demographic patterns.

It is equally important to understand that uneven development is an outcome of historical and contemporary injustice that creates and perpetuates structural inequality. In the history of urban development, urban trends and structural inequality are integrated. Suburbanization is one example. Suburbanization began in the nineteenth century as a response to conditions in industrial cities. However, given the costs of transportation to the city center, only wealthy residents initially could afford to move out of the urban core. As trolley systems and mass transit made travel more affordable, more people had the flexibility to choose where they wanted to live and many people opted to move outside the urban core to single-family residential subdivisions. Beginning in the 1930s, the federal government enacted mortgage insurance programs that facilitated suburban homeownership for many people (discussed in chapter 2 in more detail).

While people of different incomes and ethnicities sought quality affordable housing and the opportunity for homeownership, racial discrimination in housing and job markets constrained where many residents lived. Moreover, federal mortgage insurance programs institutionalized racial discrimination and devalued property in the urban core by encouraging investment in the suburbs. Differing property values impact daily life for residents in many ways. Property taxes are the basis of school funding as well as generating municipal revenues that pay for urban services and amenities. Therefore, areas with lower property values have less school funding and poorer quality urban services. In this way, the social trend of suburbanization created and intensified inequality which continues to be reflected in urban landscapes and property values today. And while housing discrimination in the US was prohibited in 1968, fair housing audits throughout the country find that housing discrimination continues (GNOFHAC 2014).

Although suburbanization continues, the "back to the city" movement also has become a highly visible social trend as white middle-income residents choose to live in central city neighborhoods rather than the suburbs (Piiparinen 2013). This is a result of many factors including a growing attraction among young professionals and baby-boomers for an urban lifestyle defined around denser places with a high level of consumption opportunities, and a rejection of the suburbs. This demographic influx has led to the gentrification of many urban neighborhoods and has been exacerbated as many cities attempt to capitalize on the trend by making their urban core appealing to these groups. The upshot is increasing property values and income for cities, safer and cleaner downtown streets, and more amenities in central city neighborhoods. However, gentrification also often results in the displacement of lower-income residents.

While increased property values and a mix of incomes can be positive, they can also hurt longtime residents if they can no longer afford to live in an area, or if they are physically or even culturally displaced when new residents with different ways of living occupy a neighborhood. For example, gentrifying neighborhoods often experience conflicts over noise and street activities when newcomers expect quiet streets that reflect the more suburban areas from where they moved.

Urban revitalization initiatives are exciting and complex. They respond to social trends and work at varying scales to improve places and reduce inequity. They involve coalitions of diverse actors who have an interest in seeing their neighborhood or city thrive. However, the visions for prosperity and ideas about how to get there diverge. Revitalization coalitions also must contend with the global and national trends and actors that play a role in shaping the places in which they work. In addition, working in market economies, coalitions must be mindful that investment in one area can contribute to decline in another. The challenges are great! Luckily, they are met with a wealth of new ideas and wonderful spaces and projects created by innovative thinkers who are committed to rebuilding better—more just, sustainable, compelling, and beautiful—urban areas than at any time before.

OUTLINE OF THE BOOK

Urban Revitalization: Remaking Cities in a Changing World comprehensively and critically synthesizes the successful approaches and pressing challenges involved in urban revitalization. Following this introduction, the book is divided into three sections. Section II examines the historical causes of decline in central cities, inner-ring suburban areas and shrinking cities. Building from the discussion of urban revitalization in this chapter, we also discuss theory useful to explain the factors that shape contemporary revitalization initiatives and outcomes. Section III provides an in-depth, critical discussion of contemporary urban revitalization policies, strategies, and projects. This section also provides a rich set of case studies that contextualize key themes and strategic areas across a range of contexts including the urban core, central city neighborhoods, suburban areas, and shrinking cities. Section IV introduces students to the analytical techniques and key data sources for urban revitalization planning. Lastly, we conclude by reflecting on the current state of urban revitalization planning and the emerging challenges the field must face in the future.

REFERENCES

Avery, Robert, Kenneth Brevoort, and Glenn Canner. 2007. "The 2006 HMDA data." *Federal Reserve Bulletin*.

Clark, Terry Nichols. 2004. *The city as an entertainment machine*. Vol. 9. Netherlands, Amsterdam: Elsevier/JAI.

Ferenchik, Makr. 2015. "Teamwork of businesses, nonprofits renews Cleveland neighborhood." *Columbus Dispatch* 26 January. www.dispatch.com/content/stories/local/2015/01/26/teamwork-of-businesses-nonprofits-hailed-as-success-in-renewing-cleveland-neighborhood.html

GNOFHAC. 2014. "Where opportunity knocks the doors are locked." Greater New Orleans Fair Housing Action Center. www.gnofairhousing.org/wp-content/uploads/2014/11/11-06-14-Where-Opp-Knocks-FINAL.pdf

Harvey, David. 1985. *The urbanization of capital, studies in the history and theory of capitalist urbanization*. Baltimore, MD: Johns Hopkins University Press.

Harvey, David. 1989. *The condition of postmodernity*. Vol. 14. Oxford: Blackwell.

Lefebvre, Henri. 1991. *The production of space*. Vol. 142. Oxford: Blackwell.

Logan, John R and Harvey L Molotch. 1987. *Urban fortunes: the political economy of place*. Berkeley, CA: University of California Press.

Molotch, Harvey. 1976. "The city as a growth machine: toward a political economy of place." *American Journal of Sociology* 82(2): 309–332.

Piiparinen, Richey. 2013. "The persistence of failed history: 'white infill' as the new 'white flight'". *New Geography* 10 July. www.newgeography.com/content/003812-the-persistence-failed-history-white-infill-new-white-flight

Schumpeter, Joseph A. 1950. *Capitalism, socialism, and democracy.* 3d Ed. New York: Harper.

Stone, Clarence N. 1989. *Regime politics: governing Atlanta, 1946–1988.* Lawrence, KS: University Press of Kansas.

Stone, Clarence N. 2005. "Looking back to look forward: reflections on urban regime analysis." *Urban Affairs Review* 40(3): 309–341. doi: 10.1177/1078087404270646.

Tallon, Andrew. 2013. *Urban regeneration in the UK.* Abingdon, Oxon: Routledge.

2

The uneven landscapes of urban development in the US

INTRODUCTION

Contemporary cities, regions, and neighborhoods are landscapes of interwoven histories that both retain the past, and reflect decades or centuries of change. In the US, colonial and early American neighborhoods can be layered with industrial landscapes in the historic cores. Similarly, patterns of early suburban development and the structure of metropolitan areas continue to impact development patterns. Suburban development along with deindustrialization also contributed to the decline in central cities after World War II. While these forces drove people from some cities in the decades following the war, in the 1980s the globalization and restructuring of urban economies brought people back to the urban core, and city centers experienced gentrification. Throughout these times, new forces of change helped to bring important improvements in urban life. In many instances, however, urban growth has resulted in population loss and local disinvestment for some places even as the overall population seems to prosper. Throughout, people have worked to revitalize their communities in the face of uneven development and they too make a mark on the urban landscape.

The purpose of this chapter is to highlight some of the major historical trends in urban development and policy that have shaped our contemporary urban landscapes. We do not set out to exhaustively chronicle US urban history. Rather, we seek to historically contextualize the processes of uneven development and revitalization that have molded the cities we live in today. We use these major trends to illustrate the layered, changing context of urban revitalization and the particular histories that comprise contemporary built environments.

Dramatic changes occurred repeatedly. Indigenous peoples lived throughout North America, including in urban settlements such as Pueblo Bonito in New Mexico. These settlements responded to both natural determinants and societal ideas about how to organize life. Colonists also settled in ways that took into account natural conditions and societal ideals. Colonial landscapes gave way to generations of economic restructuring that simultaneously rendered historic built environments obsolete and created new economic needs. Throughout, architects, engineers, developers, reformers, entrepreneurs, and regular people all influenced and shaped the built environment and, in the process, urban settlements became less responsive to natural conditions.

Each urban form and building type was designed or planned as diverse people envisioned the best way to provide housing and commercial space to meet changing needs. In all cases, different solutions were possible (as we can see by comparing urban and architectural forms around the US and the world). However, there are many commonalities because many different regions faced the challenges of density, industrialization, and globalization. People and businesses created and responded to changing circumstances, and the social trends reflected people's opportunities and migration decisions, and favored solutions for housing, streets and infrastructure, transportation, public spaces, industry, and commerce. Despite profound changes from one decade to the next, the built environment is durable. Each generation's development creates some degree of path dependency, and most subsequent changes occur incrementally. The built environment therefore changes more slowly than economic and social trends. The built environment from one generation provides rebuilding material for the next.

The processes of urban growth, change, and decline can be seen as both continuous and disruptive. Prior to European colonization of what is now the US, Cahokia, located near present-day St. Louis, developed into an expansive metropolis, but it had gone into decline and was abandoned by the time European colonists arrived. In Mesoamerica, many Mayan cities lost population and were abandoned in part due to local environmental decline brought about by urban development. Throughout the Americas, many peoples were living in North America in the fifteenth century when European colonization began, and colonization caused displacement and destruction for many people. This brings forward an important observation: urban landscapes are contested. They reflect opportunities for some people and disadvantage or, at times, destruction for others. Nevertheless, people who have faced destructive forces and discrimination, including Native Americans, African Americans, Chicanos, and the ethnically and racially diverse people who immigrated to the US throughout the centuries, continue to live in urban areas. The legacies of destruction and discrimination also continue and, as a result, contemporary revitalization objectives must work to counter these patterns of injustice. Throughout this chapter, consider that all urban development is political and the landscapes reflect how some people benefitted at the expense of others. Dismantling policies and social structures that produce inequitable development is critical to creating just and sustainable regions.

THE HISTORIC URBAN CORE

The imprints of colonial and early American cities are often-overlooked elements of contemporary urban landscapes. The walkable street grid of the urban core in many cities reflects a colonial past and shapes the dense character of central neighborhoods. The organic form of the tip of Manhattan is a remnant of its founding as New Amsterdam. The original city commons in cities such as Boston and Savannah continue to be important public spaces, and Savannah's historic core is now a national historic landmark district. The 1812 map of Savannah shows one of the first planned cities when it was founded in 1733 (Figure 2.1). Historic residential architecture defines cities' unique character too. No other city in the US, for example, has New Orleans' creole cottage and shotgun vernacular (Figure 2.2).

For many people, the warehouse districts on the edge of many contemporary downtowns are the most visible revitalized urban neighborhoods. It is easy to forget that these areas used to be crowded, mixed-use spaces of nineteenth-century manufacturing and industry, and comprised an important hub of the urban economy. Partly because walking was the primary form of local travel, cities were dense agglomerations of business, manufacturing and living quarters. In fact, few cities were larger than 3 square miles. Urban warehouse districts, in both their location and their form, reflected the conditions of that time. They located near water because it was a power source, a mode of transportation, and a resource. Rivers and other bodies of water were also a place to dispose of waste. In the factories and warehouses, workers spun yarn and processed meat. They made fabric, pots and pans, and heavy machinery. These activities took place in or near city centers, and both industrialists and workers lived near the factory.

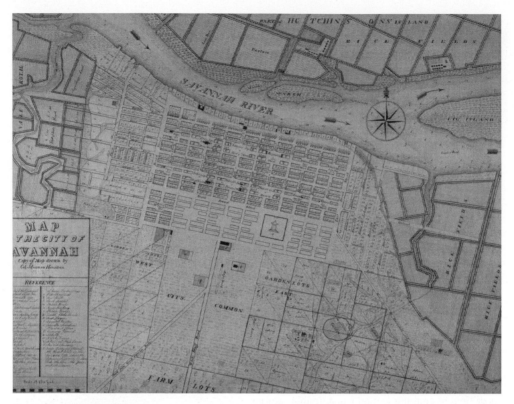

Figure 2.1   Savannah, GA 1812

Courtesy of Hargrett Rare Book and Manuscript Library

Figure 2.2   Three shotgun doubles—a distinctive New Orleans housing type

Photograph by Renia Ehrenfeucht

As industrial operations continued to evolve, they fueled new patterns of urban expansion. Industry quickly outgrew the urban warehouses as technological and managerial innovations increased the scale of manufacturing and new industries emerged. This allowed for more intensive production and increased the need for larger lots and buildings. Better transportation infrastructure, including paved streets, railroads, and intra-urban trolleys, increased the area that could be developed. Steam power and electricity made industries less dependent on nearby rivers. New industries emerged, establishing their own manufacturing districts. As a result of these changes, more space was open for development and, at the same time, some buildings and neighborhoods became obsolete.

These changes led to patterns of new investment as well as disinvestment. The shifting geography of manufacturing was also a product of changing demand for office and retail space in the urban core. Just as manufacturing industries sought out new locations outside the urban core, innovations in building technology such as fireproofed metal frame construction and the elevator allowed buildings to be both taller and more durable. It became increasingly possible for business services to locate away from sites of production in denser urban core locations. This had the effect of driving up rents in the center and further pushing both industry and housing development to the fringe. Uneven development became apparent in the industrial city as manufacturing outgrew its early warehouses and left the urban core. In some cases these warehouses were demolished and rebuilt while other buildings sat empty or were repurposed.

Paradoxically, many historic structures and districts remain because the built environment is durable and demolition is expensive, and because property owners held on to real estate, speculating that land values would again rise. Because new uses were not immediately found, some areas became neglected. This reduced their value, but it also preserved historic sites because it was less profitable to demolish them and rebuild than to build in a new area. Many of these depreciated properties contained unique historical structures and became profitable to rebuild as new trends brought people back to the urban core. These were active processes of change, where stakeholders continued to envision ways to improve urban life, but they occurred unevenly as different places at different moments became profitable and desirable.

THE ROOTS OF REVITALIZATION: LOCAL RESPONSES TO CREATE LIVABLE CITIES

Agglomeration was cost effective for businesses that relied on one another, but also it created concentrations of waste, air pollution, and noise. For most people, living conditions in the nineteenth-century city were poor and unhealthy. The basic services we take for granted had to be developed and they came about incrementally. Most nineteenth-century streets were unpaved and horses were a common form of transportation. Without trash collection, garbage was thrown into the street or public spaces to be eaten by roaming pigs and dogs. Industrial waste poured into the same sources that were used for drinking water.

From the mid-nineteenth century to World War II, Chicago's Back of the Yards neighborhood, for example, was a major global meatpacking district and the largest in the US. The neighborhood housed the Union Stock Yards, established in 1865, and the area was annexed into the city of Chicago in 1889. By 1921, the Union Stock Yards employed 40,000 people (Chicago Historical Society n.d). It was a highly efficient industrial operation. Pigs were brought in by train to be slaughtered and processed. Meat was distributed and waste from the processing (as well as unsold meat) was made into fertilizer. By contemporary accounts, the neighborhood was defined by its stench and the processing defined by the squeal of the pigs going to slaughter. Meat processing was also dangerous and dirty work. The Stock Yards inspired Upton Sinclair's famous 1906 novel *The Jungle*, which chronicled the harsh working conditions (Sinclair 1971).

As was normal at the time, the Back of the Yards neighborhood was also residential. Even as the industrialists moved away, workers still lived within walking distance, buying the modest houses when they could afford them and renting when they could not. Women shopped in the local stores that served as points of exchange for both goods and news, and men frequented the bars that served as social centers (Slayton 1986). Because of the Union Stock Yards, Chicago became the pork producer for the world and tourists would come to watch the slaughter, fascinated and horrified by the blood and screams. — Ewww!

BOX 2.1 THE BACK OF THE YARDS TODAY

Although many warehouse districts located near the downtown core have been converted into residential lofts, other former industrial districts remain empty and underused. The Back of the Yards neighborhood now suffers from underemployment and unused industrial buildings. Residents are envisioning new uses for the industrial buildings to bring much-needed jobs to the area. The Plant, a former meat processing plant, is being repurposed into a new food manufacturing center that reflects the priorities of the twenty-first-century city.

Taking advantage of its solid construction, which includes adequate drainage for food production and weight-bearing upper floors to support the weight of commercial ovens, The Plant is being transformed into an energy-independent urban farm, commercial kitchen, brewery, and small food-business incubator. The project provides technical assistance and low-cost access to assist new entrepreneurs entering the food industry. It has a commercial kitchen space for food truck operators or other food vendors, an urban farm where local businesses can grow produce for their restaurants or food trucks, and commercial microgreen, shrimp, and mushroom production to sell to local groceries.

Reflecting contemporary concerns about solid waste and energy production, the site will collect food waste from around Chicago to generate energy to power its operations. To further its internal energy efficiency, it has attracted a brewery to the site which will generate waste heat that will be used to heat the greenhouses. The Plant is using a site from the past to respond to some of today's most pressing issues: job creation, regional food production, and renewable energy.

Source: The Plant, www.plantchicago.com

Despite the difficult conditions, places like the Union Stock Yards created economic growth and jobs in cities. In this sense, the stock yards were a source of opportunity, but also created the need for revitalization. Social reformers pressured industrialists to improve working conditions and, in this way, began to create new visions for a livable city. As is the case now, historic revitalization efforts responded to both global and regional forces while improving local environments and life quality. Urban residents experimented with how to live in cities and how to solve the problems brought about by industrial urbanization.

While one response was to increasingly separate residential areas from industry, as people moved to growing and densifying cities, the municipal government also became an important avenue for people to address the challenges that arose from denser environments. Private developers and architects developed housing types to provide living spaces for the thousands of residents moving to cities. New York's apartments and tenements, Baltimore's row houses, and New Orleans' creole cottages all provided housing for many households in tight spaces. To make density livable, urban residents developed infrastructure including sewerage and water systems, improved streets, and garbage pick-up. When cities were smaller there was less of a need for collective services. In the case of garbage collection, roaming pigs and dogs ate garbage on town streets and people fed garbage to their animals. Farmers collected garbage to use as fertilizer and, in the process, removed trash from the city. As cities became dense and complex, urban residents bestowed more power on municipal governments to tax and provide services such as garbage collection, water and sewerage systems, and other public infrastructure. As urban residents recognized their collective interest in ensuring that water was drinkable and epidemic diseases could be contained, they were more likely to realize these through government action (Scobey 2002). In this way, the local government became an important participant in managing urban life. The municipal government continues to participate in revitalization efforts, in part through improving local infrastructure and services. As we discuss in chapter 9, in some situations services decline because of strained municipal revenues, and municipal agencies must develop new strategies to provide high quality services.

A movement to build urban parks and bring trees into the city was another nineteenth-century social reform initiative. Frederick Law Olmsted, his partner Calvert Vaux, and others also envisioned ways to make cities greener and offer an antidote to the problems of industrialization. Central Park and Prospect Park in New York City are celebrated examples. Boston's Emerald Necklace, Detroit's Belle Isle, and Worcester's Elm Park—the country's first municipal park—among dozens of others were also envisioned to provide lungs for the city. Tree-lined parkways also intended to provide relaxation and visual relief. These early greening initiatives responded to the dense, unhealthy conditions of the industrial city as well as providing spaces for public life, two ongoing revitalization objectives that we discuss in chapters 10 and 11.

A NEW DEAL: FEDERAL INVOLVEMENT IN URBAN DEVELOPMENT

By the 1920s, metropolitan landscapes reflected social differences and uneven access to parks and good infrastructure, as well as different house size and quality. At this time, suburban areas began to grow more rapidly than the urban core. Many people valued the residential character of suburban areas and their separation from the city's industry and diversity. Residents in wealthy, white suburbs directed their resources to building good neighborhoods for themselves that excluded others, but people from different communities also sought to buy houses or property. Recent immigrants, African American households, and other people of color valued homeownership despite significant discrimination that resulted in fewer neighborhoods to which they could move, and significantly higher housing costs. Working class people—even in neighborhoods that might appear undesirable, such as the Back of the Yards—would go through great sacrifices to buy property. People of all incomes and ethnicities saw homeownership as a means to greater independence and security as well as a way to gain status through wealth accumulation (Kruse and Sugrue 2006, Nicolaides 2002, Roediger 2005, Wiese 2004).

In 1929, the stock market crash and the subsequent Great Depression altered housing and work opportunities, and created problems that cities could not deal with alone. Although opportunities for homeownership were threatened by the financial collapse, in the 1930s the federal government became involved. Roosevelt's New Deal turned to urban and regional development as a solution to temporary unemployment. In doing so, the federal government extensively involved itself in constructing local infrastructure and public buildings as well as stabilizing and expanding housing markets. Federal policies became an important influence on patterns of urban investment and disinvestment throughout the rest of the twentieth century.

One impact on cities and regions was the redevelopment and maintenance performed by people hired through the numerous work programs. Urban and regional revitalization occurred on a large scale through the efforts of the Civilian Conservation Corps, the Public Works Administration, the Civil Works Administration, the Works Progress Administration (WPA), and the Tennessee Valley Authority. Each put men and women to work building and repairing infrastructure and public buildings, building schools and parks, improving national parks, and increasing flood protection. The WPA alone built over 6,000 schools and renovated, expanded and built over 2,500 hospitals. The WPA also commissioned artists and writers who produced murals, theater, and other art, including this WPA mural in the US courthouse in Albuquerque (Figure 2.3) as well as documenting American life and culture. It also constructed 1,668 parks and improved over 6,500 more (Leighninger 2007). Virtually every city has a legacy of this tremendous investment in public works reflected today in its parks, bridges, and public buildings.

Perhaps the most widespread impact, however, was on housing. While the ideal of suburban life came from people living in the dense confines of the industrial city, changes to residential lending made homeownership affordable to more residents. The federal government had privileged homeownership since 1913, when it made mortgage interest deductible from the newly established federal income tax (Leighninger 2007). It took a much bigger step, however, under the Federal Housing Act of 1934. The Act created the Federal Housing Administration (FHA) to insure new home loans and, in exchange, required banks to allow longer repayment periods and offer lower interest rates. It also allowed prospective home buyers to borrow a higher percentage of the appraised value of their property, which reduced the amount necessary for a down payment. Because banks knew that if the property owner defaulted they would not lose their capital, they were willing to make long-term, low-interest home loans. Following the passage of the FHA in 1934, new housing starts accelerated from 93,000 in 1933 to 619,000 just eight years later (Jackson 1986: 205).

Despite its achievements in increasing homeownership opportunities, the FHA also contributed to uneven urban development. The agency allowed racial discrimination through guidelines that outlined the areas where the federal government would provide mortgage insurance. The FHA insured properties with high appraisal value and used eight factors to calculate the appraisals, including relative economic stability and protection from adverse influence, one of which was racial and ethnic diversity. In its 1935 underwriting manual, the FHA listed "infiltration of inharmonious racial or nationality groups" as one potential adverse influence (quoted in Schwartz 2010: 55).

The biases in the risk assessment protocols reinforced a discriminatory lending process commonly referred to as redlining. Redlining emerged with the national Home Owners Loan Corporation (HOLC), which was created in 1933 to refinance mortgages in danger of default (Jackson 1986). The HOLC developed an appraisal system based on four color-coded categories of lending risk, and mapped entire cities according to these categories. Those in the highest risk categories were colored red. Areas with older building stock, mixed uses, and high percentages of people of color, particularly African American residents, or racially mixed neighborhoods were considered the highest risk. These guidelines facilitated segregation and they also contributed to the situation that when people of color moved to an area, property values would decline because the areas became higher risk by their presence. Even though people of all incomes sought opportunities for homeownership, the HOLC and FHA contributed to discrimination in the real estate industry.

The programs also privileged homogenous new construction, and this stimulated a trend in which middle- and upper-income residents moved out of the central city and into new, segregated suburbs. The programs promoted suburbanization by devaluing older building stock and mixed uses, which facilitated central city decline. For instance, between 1935 and 1939, in St. Louis, over 90% of households buying mortgage insurance were in the suburbs (Jackson 1986). This lending system contributed significantly to the uneven patterns of development and opportunity that dominated the US since World War II.

Figure 2.3   A WPA mural in Albuquerque's US courthouse
Photograph by Carol M. Highsmith

POLICY AND MARKETS: SUBURBANIZATION AND CENTRAL CITY DECLINE AFTER WORLD WAR II

Suburbanization and central city decline were interrelated trends that accelerated in the decades that followed World War II. After World War II ended, the mortgage insurance programs that began in the New Deal were expanded. The GI bill enabled many returning veterans to buy a house—and with the acute housing shortages in the cities, the new suburbs were where they landed. The FHA made it cheaper to buy a home rather than rent one, and since FHA standards favored suburban single-family homeownership, it also meant that living in the city center made less economic sense than a suburban location (Jackson 1986). More people than at any time previously became homeowners and spread urban development around the metropolitan landscape (Jackson 1986).

The growing infusion of capital in the suburbs combined with the suburban bias in federal policy caused capital available to invest in urban neighborhoods to disappear. State and federal highway funding and the development of the Interstate Highway System beginning in the 1950s further drew people and money from central city areas (Teaford 1990). In conjunction, like the residential population, jobs and industry moved further outward. This process is often referred to as deindustrialization because cities that formerly had massive employment in heavy manufacturing industries, such as Detroit, Cleveland, Chicago, and New York, experienced a loss of their industrial base to suburban municipalities at first and, later, overseas. As more industries moved out of the urban core during and after World War II, housing discrimination trapped many people of color in job-poor, declining central city neighborhoods.

The relationship between suburbanization and central city decline was quickly visible. By the 1940s, federal housing policies began to focus on the worsening conditions in the urban core. The Housing Act of 1949 attempted to stimulate urban renewal, as we discuss in chapter 3, by redeveloping areas that suffered from disinvestment. In this way, federal housing policies may be viewed as part of a larger policy process focused on stimulating economic growth through the simultaneous decentralization or suburbanization of metropolitan areas *and* the recentralization and revitalization of city centers. However, this had disparate impacts on people. Urban renewal initiatives often displaced many of the residents who lived in the urban core, and even though civil rights activists demanded that African Americans also received GI benefits and the end to discriminatory lending, people of color disproportionately remained in the declining central city. Viewed together, suburbanization and urban renewal allowed continued economic expansion in the US following World War II while further entrenching inequality and hurting many people who lived in the central city.

By the 1980s, many suburban municipalities transformed from bedroom communities to edge cities defined by a concentration of jobs and businesses (Garreau 1991, Teaford 1997). Suburban job growth and the stronger revenue stream associated with it made suburban residents less dependent on and invested in the central city. As a result, the urban core fell into deeper decline. Instead of seeing a shared fate, many suburban municipalities feared that the decline would spread and strategic regional action became difficult to accomplish. David Rusk, the former mayor of Albuquerque, has argued that in order to strengthen US urban areas, cities and suburbs cannot be isolated, and cities that annexed the outlying areas have fared better (Rusk 2013). Although Rusk and others have advocated for consolidating urban regions, this has been difficult to accomplish.

A GLOBALIZED WORLD: DEINDUSTRIALIZATION, ECONOMIC RESTRUCTURING, AND GENTRIFICATION

By the 1970s and 1980s, it became clear that a larger process of deindustrialization was underway. As we discuss in more depth in chapter 4, the mobility of manufacturing continued to impact cities and urban residents, particularly in what became known as the Rust Belt. Between 1963 and 1989, East and Central North regions experienced a 30% drop in industrial output, resulting in declining income and job opportunities. During this time, industry increased by 33% in the South and the West (Crandall 1993). By the 1980s, manufacturing became more footloose and functions were dispersed around the globe. During the 1960s, the economic growth in the US averaged 4.1% annually. Over the decade the average family had a third more spendable income. Between 1973 and 1980, there was no income gain (Bluestone and Harrison 1982). The effect on former industrial cities was the most profound. They lost jobs, residential population, and their tax base.

Many cities continue to experience population and job loss, as we discuss further in chapter 9. In the 1980s, however, some central cities began to show signs of a revival. Cities including New York, Paris, and Tokyo among others emerged as global financial hubs, and others such as Miami and Toronto became regional hubs in an increasingly globalized economy driven by business and knowledge-based services rather than the production of material goods (Sassen 2012).

The rise of the financial services and technology sectors entailed the growth of a sizable workforce with high discretionary spending. The rapid growth of business services and office development in the central business districts began to generate demand by middle-class professionals for central city neighborhoods. Blighted areas and abandoned buildings became home to new residential populations and new businesses. New homeowners, retail and entertainment outfits, and arts groups began to renovate historic buildings. With this came rising property values, lower vacancy rates, and new sources of much-needed revenue for cities.

Early on, the move back to the city center reflected demand for spaces and opportunities not provided in the suburbs such as warehouse spaces where artists could work, but it quickly became a much wider trend capitalized on by real estate speculators (Ley 1996, Zukin 1982). Changing demographic and lifestyle preferences also helped explain the interest in downtown. Increasingly, young adults and "empty nesters" gravitated to urban centers. Downtown boosters argued that young people in particular were rejecting a suburban way of life defined around a single-family home, a long commute, and chain stores because they favor walkable, urban settings (Leinberger 2009). Some have argued that a new, mobile "creative class," comprised of people working in science and technology, business, entertainment, and other knowledge-based industries, elect where to live based on quality of life, amenities, and a diverse, urban environment (Florida 2002). To attract these new residents, downtown leaders promote amenities including their historic architecture, industrial spaces, and dense and pedestrian-friendly street grid. These latent assets became more valuable as economic and social trends changed.

The trend toward urban living grew dramatically in the twenty-first century. According to a study by the Brookings Institution, in 2010, the population growth in a region's primary city grew at a faster rate than suburbs in 27 of the largest 51 metro areas for the first time since 1920 (Frey 2012). Much of this growth has been driven by young, upwardly mobile individuals. As another recent study found, the number of college-educated 25–34-year-olds increased twice as quickly in downtowns and neighborhoods within 3 miles of the urban center than in the rest of the region (El Nasser 2011). This demographic comprised almost a quarter of the US downtown population, and 44% of all downtown residents held a bachelor's degree or higher in 2000 (Birch 2005).

These changes signal what is now commonly labeled as gentrification, or the reinvestment in and rehabilitation of central city industrial areas and residential districts by developers and more affluent households (Lees, Slater, and Wyly 2007, Smith 1996). Some explain gentrification as a result of changing economic conditions and stress a production-side theory (Smith 1979, 1996). According to Neil Smith, the central causal mechanism of gentrification is the existence of a "rent gap"—the difference between the current and potential value of a central city property under a new land use. In brief, decades of disinvestment driven by suburbanization and deindustrialization created an economic opportunity for investors. The rent gap emphasizes how uneven urban development drives the gentrification process.

Others make a consumption-side argument focusing on the importance of the cultural and symbolic dimensions of gentrification (Ley 2003, Zukin 1982). In this view, gentrification is the result of changing tastes and social trends. Some people have found suburban life too homogenous and turned to urban centers for their diversity, historic architecture, and amenities. In this way, central city living becomes a means of expressing one's identity. For example, as many document, the first wave of gentrification may occur as artists inhabit neighborhoods perceived as blighted. They set the stage for neighborhood change by rehabilitating and aesthetically revaluing aging industrial, residential, and commercial buildings into a "neo-bohemia" of art studios, galleries, bars, coffee shops, and restaurants. Their efforts serve to change the look and feel of urban neighborhoods, which attracts higher-income groups (Deutsche and Ryan 1984, Ley 2003, Lloyd 2010, Mathews 2010, Zukin 1982, 2010). Moreover, cities have capitalized on this trend by building more spaces of consumption that instigated new waves of gentrification (Ley 2003, Zukin 1982, 2010).

Gentrification is a complex phenomenon that has multiple outcomes in different places. Gentrified neighborhoods and downtown entertainment districts particularly in major financial centers such as London, New York, San Francisco, and Toronto helped to push central city rents upward. This in turn led to the displacement of households and businesses that could not afford to remain in the area. In particular, the gentrification of urban core neighborhoods squeezed poor African American and Latino residents already suffering from job losses due to urban restructuring. The neighborhood improvements that came with the reinvestment increased demand to a point that raised rents, property values, and property taxes beyond the reach of some longtime homeowners, renters, and businesses. In such instances, maintaining affordability while increasing neighborhood opportunity without displacement becomes a challenge (Bridge, Butler, and Lees 2011, Newman and Wyly 2006).

Nevertheless, in other cases, place-based improvements benefitted residents in the area (Byrne 2002, Duany 2001). In some circumstances, neighborhood change may reflect normal rates of residential turnover among disadvantaged and minority households rather than direct displacement (Freeman 2005, Freeman and Braconi 2004, McKinnish, Walsh, and White 2010). Further, some cities desperate for new investment may welcome gentrification as a route to revitalization, arguing that the surplus of vacant property is so great that displacement can be controlled. Gentrification has become a widespread topic as people express a desire for investment without displacement and for the arrival of new services, as the satirical newspaper *The Onion* portrays (Box 2.2). However, even Detroit has experienced gentrification and displacement in select areas where costs have substantially increased (Reindl and Gallager 2014).

BOX 2.2 DECAYING CITY JUST WANTS TO SKIP TO PART
WHERE IT GETS REVITALIZED RESTAURANT
SCENE

Camden, NJ—Saying they were fed up with the numerous challenges stemming from their city's extensive urban decay, Camden, NJ residents confirmed Wednesday that they would love to just skip to the part where they get a hip, revitalized restaurant scene. "I realize that these boarded-up storefronts and abandoned factories might be turned into trendy cafés and bistros someday down the line, but I think most of us would be pretty thrilled if we just went ahead and got to that stage right now," said resident George Pierson, noting that he is fully willing to bypass Camden's endemic crime, rampant drug abuse, and high unemployment rate in order to jump right to the point when he and the city's other occupants can enjoy dozens of farm-to-table gastropubs. "Sure, we'll eventually see lobster roll stands and high-end noodle bars popping up on every corner, but that could take years or even decades. Let's just skim over all the gang turf disputes and burnt-out streetlights and go straight to blocks lined with stores specializing in key lime pies, locally sourced butcher shops, and gourmet empanada places. That honestly seems like the way to go." Camden residents also told reporters they would like the city's accelerated revitalization process to then stop just before they are priced out of their current apartments.

Source: *The Onion* (2015)

Even in circumstances where longtime residents are not physically displaced, they may suffer from cultural displacement if new residents with different ways of living occupy a neighborhood. New residents may bring to a neighborhood social and cultural changes that create tensions between old and new residents. While some improvements such as new parks, street lighting, and grocery stores are welcomed by all, others such as high-end boutiques, restaurants, and art walks may send a message of exclusion (Shaw and Sullivan 2011, Zukin 2010). Further, gentrifying neighborhoods often experience conflicts over noise and street activities when newcomers expect quiet streets that resemble the more suburban areas from where they moved.

For these reasons, it is important not to assume that gentrification or physical improvements in a neighborhood indicate revitalization. It is critical to consider the impacts on the people who live there. Local governments and other actors in the redevelopment process have long helped to facilitate the gentrification of urban neighborhoods (Hackworth and Smith 2001). As we discuss in chapter 3, the federal government played a role in gentrification through the urban renewal program by funding the demolition of established urban neighborhoods so that redevelopment agencies could replace them with high-end residential and entertainment complexes. More recently, efforts to promote mixed-income communities in high-poverty areas through the housing program HOPE VI has contributed to gentrification and displacement as well (Bridge, Butler, and Lees 2011, Goetz 2013). Local governments play a part in gentrifying the urban core through infrastructure investment, tax increment financing (TIF) districts, and other development incentives that we cover in chapter 6. However, as we also discuss, the public sector can intervene in market processes to balance the detrimental outcomes of uneven development.

SUMMARY

In just 150 years, US cities grew from small compact centers to sprawling metropolises. Even though the dense industrializing cities of the nineteenth century rarely exceeded 3 square miles, by the twenty-first century these cities had grown into multi-centered regions with miles of suburbs and dozens of municipalities. Los Angeles metropolitan area, for example, now has a land area of close to 5,000 square miles. Contemporary Los Angeles still has a central plaza and warehouses along with densely packed blocks in the central core that were once ringed with early suburbs. As the city expanded outwards, the urban core annexed the early suburbs and newer suburban development extended for miles. In the process, twentieth-century urbanism became characterized by single-family residential landscapes dotted with multiple commercial centers accessible by a network of freeways.

Changing expectations and needs can lead to shifts at different scales—from one building to another, from a central city to its suburbs, and from one region or country to another. As a result, one generation's vision can become a problem for the next. Conversely, a problem at one moment can become a new opportunity as economies and social trends change and people find new uses for obsolete structures. The process by which land is valued, devalued, and revalued in a market system, what Harvey labelled a "spatial fix," forms the basis for future development and revitalization. Buildings and sometimes even whole neighborhoods go into decline and are subsequently abandoned. They might be demolished or converted to a different use. New development has different form, responding to the ideas and needs visible at the time. The cycles of investment, disinvestment, and reinvestment continue to unfold unevenly.

Even though buildings or a given street might no longer work well, myriad components of the built environment, from property divisions and ownership to infrastructure and buildings, persist long after their primary purpose changes. When local governments and community groups pursue strategies for urban revitalization, they work within and respond to the changing and conflicted urban landscape. They also work with the diverse people who call the area home.

Every phase of urbanization, from the industrial city and the early suburbs to downtown urban renewal and the gentrification of central city neighborhoods, is driven in part by shifting investment among places. It is also driven by the actions people take to capture that investment, mitigate its adverse impacts, or transform the area in response to new concerns and values. These cumulatively result in the varied urban landscapes across the US.

STUDY QUESTIONS

1   Identify the historic neighborhoods in your region. Visit a neighborhood and take note of the block length and width, the lot size, the scale, and the character of the neighborhood. Next identify the suburban neighborhoods in your region. Visit a suburban neighborhood (within or outside the city). Compare the neighborhood form of the two areas. Hint: you can often see the difference in early and later development by looking at street maps of the area. How would you describe the neighborhoods today? Are they different from or similar to their historic uses?
2   How did New Deal housing policies affect the urbanization of US cities? Can you see the legacy of this development pattern in your region?
3   Are there neighborhoods in your city that you consider gentrified? Do you consider these areas as good examples of urban revitalization? Why, or why not?

REFERENCES

Birch, Eugenie L. 2005. *Who lives downtown?* Living Cities Census Series. Washington, DC: Brookings Institution.

Bluestone, Barry and Bennett Harrison. 1982. *The deindustrialization of America: plant closings, community abandonment, and the dismantling of basic industry.* New York: Basic Books.

Bridge, Gary, Tim Butler, and Loretta Lees. 2011. *Mixed communities: gentrification by stealth?* Bristol: Policy Press.

Byrne, J Peter. 2002. "Two cheers for gentrification." *Howard Law Journal* 46: 405–432.

Chicago Historical Society. n.d. "The birth of the Chicago Union Stockyards." Accessed 18 March. www.chicagohs.org/history/stockyard/stock1.html

Crandall, Robert W. 1993. *Manufacturing on the move.* Washington, DC: Brookings Institution.

Deutsche, Rosalyn and Cara Gendel Ryan. 1984. "The fine art of gentrification." *October* 31: 91–111.

Duany, Andres. 2001. "Three cheers for 'gentrification'." *American Enterprise* 12(3): 36–39.

El Nasser, Haya. 2011. "Urban centers draw more young, educated adults." *USA Today* 4 January. http://usatoday30.usatoday.com/news/nation/2011-04-01-1Ayoungrestless01_ST_N.htm

Florida, Richard L. 2002. *The rise of the creative class: and how it's transforming work, leisure and everyday life.* New York: Basic Books.

Freeman, Lance. 2005. "Displacement or succession? Residential mobility in gentrifying neighborhoods." *Urban Affairs Review* 40(4): 463–491. doi: 10.1177/1078087404273341

Freeman, Lance and Frank Braconi. 2004. "Gentrification and displacement: New York City in the 1990s." *Journal of the American Planning Association* 70(1): 39–52. doi: 10.1080/01944360408976337

Frey, William. 2012. "Demographic reversal: cities thrive, suburbs sputter." Washington, DC: Brookings Institution. www.brookings.edu/research/opinions/2012/06/29-cities-suburbs-frey

Garreau, Joel. 1991. *Edge city: life on the frontier.* New York: Doubleday.

Goetz, Edward G. 2013. *New Deal ruins: race, economic justice, and public housing policy.* New York: Cornell University Press.

Hackworth, Jason and Neil Smith. 2001. "The changing state of gentrification." *Tijdschrift voor economische en sociale geografie* 92(4): 464–477.

Jackson, Kenneth T. 1986. *Crabgrass frontier: the suburbanization of the United States.* New York: Oxford University Press.

Kruse, Kevin Michael and Thomas J Sugrue. 2006. *The new suburban history.* Chicago, IL: University of Chicago Press.

Lees, Loretta, Tom Slater, and Elvin Wyly. 2007. *Gentrification.* New York: Routledge.

Leinberger, Christopher B. 2009. *The option of urbanism: investing in a new American dream.* Washington, DC: Island Press.

Leighninger, Robert D. 2007. *Long-range public investment: the forgotten legacy of the New Deal.* Columbia, SC: University of South Carolina Press

Ley, David. 1996. *The new middle class and the remaking of the central city.* London: Oxford University Press.

Ley, David. 2003. "Artists, aestheticization and the field of gentrification." *Urban Studies* 40(12): 2527–2544.

Lloyd, Richard. 2010. *Neo-bohemia: art and commerce in the postindustrial city.* New York: Routledge.

Mathews, Vanessa. 2010. "Aestheticizing space: art, gentrification and the city." *Geography Compass* 4(6): 660–675.

McKinnish, Terra, Randall Walsh, and T Kirk White. 2010. "Who gentrifies low-income neighborhoods?" *Journal of Urban Economics* 67(2): 180–193.

Newman, Kathe and Elvin K Wyly. 2006. "The right to stay put, revisited: gentrification and resistance to displacement in New York City." *Urban Studies* 43(1): 23–57.

Nicolaides, Becky M. 2002. *My blue heaven: life and politics in the working-class suburbs of Los Angeles, 1920–1965.* Chicago, IL: University of Chicago Press.

The Onion. 2015. "Decaying city just wants to skip to part where it gets revitalized restaurant scene." May 13, 51 (19). www.theonion.com/article/decaying-city-just-wants-skip-part-where-it-gets-r-50409

Reindl, JC, and John Gallager. 2014. "Downtown Detroit apartment rents spiking higher, even pricing out middle class." *Detroit Free Press* May 25. www.freep.com/article/20140126/BUSINESS04/301260081/ Rents-rising-Detroit-downtown-Midtown-Corktown

Roediger, David R. 2005. *Working toward whiteness: How America's immigrants became white: the strange journey from Ellis Island to the suburbs.* New York: Basic Books.

Rusk, David. 2013. *Cities without suburbs: a census 2010 perspective.* Washington, DC: Woodrow Wilson Center Press.

Sassen, Saskia. 2012. *Cities in a world economy.* Thousand Oaks, CA: Sage.

Schwartz, Alex. 2010. *Housing policy in the United States.* New York: Routledge.

Scobey, David M. 2002. *Empire city: the making and meaning of the New York City landscape.* Philadelphia, PA: Temple University Press.

Shaw, Samuel and Daniel Monroe Sullivan. 2011. "'White night': gentrification, racial exclusion, and perceptions and participation in the arts." *City & Community* 10(3): 241–264. doi: 10.1111/j.1540-6040.2011.01373.x

Sinclair, Upton. 1971. *The Jungle*. Cambridge, MA: R. Bentley.

Slayton, Robert A. 1986. *Back of the yards: the making of a local democracy*. Chicago, IL: University of Chicago Press.

Smith, Neil. 1979. "Toward a theory of gentrification: a back to the city movement by capital, not people." *Journal of the American Planning Association* 45(4): 538–548. doi: 10.1080/01944367908977002

Smith, Neil. 1996. *The new urban frontier: gentrification and the revanchist city*. London: Routledge.

Teaford, Jon C. 1990. *The rough road to renaissance: urban revitalization in America, 1940–1985*. Baltimore, MD: Johns Hopkins University Press.

Teaford, Jon C. 1997. *Post-suburbia: government and politics in the edge cities*. Baltimore, MD: Johns Hopkins University Press.

Wiese, Andrew. 2004. *Places of their own: African American suburbanization in the twentieth century*. Chicago, IL: University of Chicago Press.

Zukin, Sharon. 1982. *Loft living: culture and capital in urban change*. Baltimore, MD: Johns Hopkins University Press.

Zukin, Sharon. 2010. *Naked city: the death and life of authentic urban places*. Oxford: Oxford University Press.

SECTION II

Local Change in a Global Economy
The History and Theory of Urban Revitalization

In section II, we examine important moments in urban revitalization policy and the key actors that make urban revitalization happen. In chapter 3, we begin by concentrating on three federal policy agendas between the 1940s and 1970s: urban renewal, the War on Poverty, and the New Federalism. Federal policies provide significant sources of funding for revitalization efforts, and federal regulations also influence how urban revitalization projects are designed and implemented. In chapter 4, we turn our attention to the development of neoliberal governance and the ways that this political framework has reshaped approaches to urban revitalization from the 1970s to the present. This discussion is set against the backdrop of urban economic restructuring. Each of these periods marks a shift in the contours of urban policy. Through this policy survey, we contextualize the historical causes of decline in different urban areas and the ways in which the federal government responded. We also see how the larger economic and political context intersects with local conditions to shape policy decisions and outcomes. In chapter 5, we focus our attention on the politics of urban development and the people involved in urban revitalization at the local level. Here we look at how an array of actors, from grassroots activists to business-oriented urban regimes, clash and collaborate to shape the development process and revitalization outcomes.

Figure II.1 presents a historical timeline of the key policies, events, and concepts that run through this section, and those discussed in chapter 2. As you read the following chapters, it will be useful to refer back to this timeline to review how the specific policies relate to different policy eras and major social and economic trends.

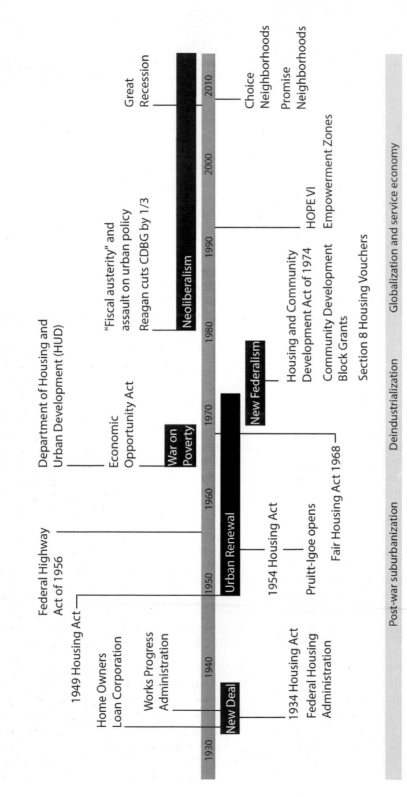

Figure II.1 Historical timeline of urban revitalization

Source: Carl Grodach

3

Urban revitalization in historical perspective

Federal urban policy, 1940s–1970s

The federal government has played an important role in enabling and shaping urban policy. Local governments have autonomy in how they craft and implement their revitalization strategies, but they are highly dependent on federal funding, funding implementation rules, and national policy objectives. As we have discussed, Roosevelt's New Deal created programs to encourage urban infrastructure and housing development as a way to address unemployment during the Great Depression. The HOLC and FHA, the Housing Act of 1937 (which initiated the first dedicated public housing program in the US), the Interstate Highway System, and later federal housing programs that we will discuss here influenced patterns of both investment and disinvestment. These programs helped to steer private development toward particular areas and away from others. Most of these programs focused on the creation of new homeownership and job opportunities in the suburbs. However, by steering investment to suburban communities, the federal government contributed to worsening housing and neighborhood conditions in many urban neighborhoods. While policymakers have recognized this contradiction and sought to respond to urban decline, they did not always assist those hardest hit by uneven development.

We focus on three distinct phases of federal urban policy from the 1940s to the 1970s.[1] In each phase, we examine how key policy decisions both responded to and contributed to patterns of growth and decline. First, we discuss the urban renewal period, which defined federal urban policy from the late 1940s to the early 1960s (and continued as an official government program until 1974). Urban renewal had many objectives in service of the public interest, including "slum clearance," the provision of adequate housing for low-income families, and better planned urban communities. However, in many cities urban renewal was infamous for guiding investment toward the demolition and reconstruction of central city business districts and neighborhoods at the behest of downtown interests. Next, we examine the War on Poverty. This phase of federal policy emerged in response to an extremely challenging set of problems including the failure of urban renewal to stave off the growing concentration of poverty and joblessness in urban neighborhoods, mounting racial tensions, and brewing urban unrest. During this period, urban policy was a component of larger efforts around job creation and building social safety net programs. It marked a brief turn from the bricks-and-mortar approach of urban renewal toward a more comprehensive and people-oriented approach to revitalization. Finally, we conclude the chapter with a discussion of the New Federalism, a policy period defined by the devolution of federal powers to state and local authorities and a move away from targeting underserved central city areas. This phase of federal policy emerged in the late 1960s and early 1970s, laying the foundation for the market-driven neoliberal urban policy discussed in chapter 4.

Between the New Deal in the 1930s and New Federalism in the 1970s, the role of the federal government in urban revitalization ebbed and flowed. Prior to the 1930s, the federal government had little involvement in urban policy. It then emerged as an active source of urban redevelopment that extended through urban renewal and, to some extent, through the War on Poverty. In the 1970s, the federal government assumed a less active role, offering assistance while giving local and state governments more autonomy. By the 1980s and 1990s, as we discuss in chapter 4, overall aid to cities dramatically declined. The federal government continues nevertheless to play an active and influential role in urban revitalization.

TABLE 3.1 PHASES OF FEDERAL URBAN POLICY, 1940s–1970s

Time period	Policy phase	Key programs and legislation	Defining features
1949–74	Urban renewal	Housing Act of 1949 Housing Act of 1954	Rebuilding downtown neighborhoods and CBD Downtown growth coalitions Displacement of poor, minority residents New public housing and business and entertainment districts
1964–69	War on Poverty	Economic Opportunity Act Community Action Program Special Impacts Program Model Cities Department of Housing and Urban Development (HUD)	Attempt to address racial inequality through developing human capital Community participation and leadership Emphasis on social services and "people-based" programs alongside rebuilding
1969–1970s	New Federalism	Housing and Community Development Act of 1974 Community Development Block Grants Section 8 Housing Vouchers Urban Development Action Grants	Local autonomy Privatization Market-based policies

UNDERSTANDING FEDERAL POLICY

Three key characteristics define federal urban policy. Each has had a lasting impact on urban revitalization practice and thinking. First, federal policy has tended to prioritize *place-based* efforts that concentrate on redeveloping the built environment. Place-based mechanisms are geared toward improving conditions in specific neighborhoods, most often by supporting efforts to rebuild housing, commercial facilities, and infrastructure in targeted areas. This approach has brought many improvements to central city neighborhoods, but the benefits are not always equally distributed. Too often, the improvements have done little to create opportunities for lower-income residents.

Second, urban revitalization has tended to prioritize market-driven strategies, which influence the geography of federal and local investment. As many urban scholars argue, federal policy has historically been oriented toward stimulating capital investment rather than specifically improving living conditions for poor residents (Fogelson 2001, Jackson 1986, O'Connor 2013). Indeed, early urban programs, including the FHA (initiated in 1934) and post-World War II urban renewal, as well as contemporary policies have all been geared toward stimulating economic growth by incentivizing private investment in targeted areas. As a result, they direct resources to those areas with strong market potential rather than struggling places that are poorly positioned for growth.

Third, the place-based and market-driven emphasis in federal urban policy has failed to address the structural causes of poverty and unemployment, which reproduce disadvantage in urban areas. Because the factors influencing inequality and urban decline emanate from beyond the local level, they are difficult for local-level policymakers to tackle without federal assistance. At the same time, when assistance is in place, federal policy has had mixed results. Project objectives, limited funding, and the structure of the programs have all undermined potential gains at one time or another.

Nevertheless, local municipalities face challenges that they do not have resources to address. Federal assistance continues to be necessary in both growing and shrinking areas to build affordable housing, offer social programs, and maintain or build infrastructure. For declining areas, or areas with temporarily weak markets, cities also need federal assistance to further their revitalization initiatives.

URBAN RENEWAL

Urban renewal emerged against the backdrop of post-World War II suburbanization and the resultant disinvestment in central cities across the country. By 1950, the suburban population was growing ten times faster than that of the city proper (Teaford 1986). As suburban areas experienced an influx of residential investment, office and industrial jobs, and new infrastructure, city centers faced divergent prospects. Conditions in urban core neighborhoods and business districts were characterized by rising unemployment, poor housing conditions, and aging infrastructure. Because suburbanization channeled investment elsewhere, older central city areas lost the tax base which further hindered their efforts to address these issues. Facing housing and job discrimination in the suburbs, people of color disproportionately continued to live in disinvested central city neighborhoods.

The federal government responded with the Housing Acts of 1949 and 1954. This legislation facilitated what has come to be known as urban renewal, possibly the most criticized federal urban revitalization policy on record. The federal government officially engaged in urban renewal because it recognized that urban blight had become a problem of national scope and that most cities did not possess the financial resources to address redevelopment on their own. The Housing Act of 1949 set out to "eliminate substandard and other inadequate housing through the clearance of slums and blighted areas ... and to provide adequate housing for urban and rural nonfarm families with incomes so low that they are not being decently housed" (US Senate 1949: 1). The application of the urban renewal program, however, was often less about improving poor neighborhoods for the people who lived there than about redeveloping downtown business districts.

As a result, we can view these federal housing policies as part of a wider government effort to stimulate national economic growth. Urban renewal policies were intended to recentralize investments in city centers after two decades of encouraging decentralized urbanization. Viewed together, suburbanization and urban renewal served as a "spatial fix" to ensure continued economic expansion in the US following World War II (Harvey 1985). Through this lens, urban renewal is thus one component of a larger, contradictory policy process that simultaneously funded *and* defunded central city development.

The key legislation enabling federal funding for urban redevelopment was Title I of the Housing Act of 1949. Title I provided federal subsidies to cover up to two-thirds of the costs that redevelopment agencies incurred investing in redevelopment projects. Cities could use funds to purchase, clear, prepare, and develop land in designated redevelopment zones, initially for "predominantly residential" use (Fogelson 2001: 377). The remaining one-third costs could come from the donation of land or the construction of new streets, sewers, and other infrastructure for a project, all of which provided inducements to attract private developers and spur downtown revitalization. Key to the redevelopment process was the use of eminent domain, which permitted the quasi-public redevelopment agencies to take private land and reimburse property owners at the current fair market value. This process enabled redevelopment agencies to acquire and completely level entire city blocks and rebuild new streets, utilities, and infrastructure in preparation for sale or lease to private developers. The properties were typically sold or leased under market value. The Housing Act of 1954 expanded this activity by shifting the emphasis from public housing to commercial development. In the process, public housing construction declined from a peak of 58,000 per year in 1952 to 24,000 in 1964, even as urban renewal funding increased (Flanagan 1997). In total, the federal government spent an estimated $30 billion (in 2000 US dollars) over the course of the program (Hyra 2012).

At first local governments and other downtown interests assumed suburbanites would commute back to downtown for work, but they quickly saw that metropolitan expansion was hollowing out and draining the urban core of resources as jobs, businesses, and retail followed residential populations to the suburban fringe (Fogelson 2001). In response, coalitions of politically connected downtown business leaders and property owners in many cities sought to protect their interests and redevelop blighted areas in their central business districts (CBDs). Downtown leaders envisioned the transformation of struggling neighborhoods into destinations that could lure middle- and upper-class suburban households back to the city. However, they faced major hurdles. Not only were real estate prices high in and around the CBD, but also large tracts of land would be necessary for large-scale redevelopment. Many areas, ignored for decades, contained a substantial number of structures in poor physical condition.

According to downtown boosters, large swaths of the CBD would require complete rebuilding and a shift in the character of land uses to attract investors. Downtown growth coalitions lobbied for federal support to realize their redevelopment plans, but attracting the wealthy was not a goal of existing federal housing and redevelopment programs. As a result, as Fogelson (2001) argues, downtown interests had to reinterpret the concept of blight from that of a social problem—in which poor neighborhood conditions were seen as causing social and moral decay—to framing blight as an economic liability. Urban renewal was championed as a program to stop the spread of blight and check the advancement of "incipient slums" that threatened to depress entire urban economies (ibid.: 349). Downtown business interests claimed that current uses reduced redevelopment prospects and public housing would not increase land values. In contrast, they argued, commercial development and upmarket housing would spur an economic renaissance by generating consumer spending and bringing businesses back to downtowns.

It was in this context that cities established the first redevelopment agencies to oversee urban renewal programs. These powerful entities were bestowed with the power to obtain, buy, prepare, and sell land for development. While officially charged with overseeing public improvements, they often worked hand in hand with local elites who dictated where and how redevelopment occurred with little to no accountability.

Initially, urban renewal found supporters among divergent groups. Downtown interests saw Title I as a means of preserving property values and attracting suburbanites while mayors and local governments viewed it as a means of removing blighted neighborhoods and boosting municipal coffers. Additionally, public housing advocates supported urban renewal as a means to construct much needed public housing. Planners and architects saw the legislation as a means to more rationally and efficiently design the city. Central city cultural institutions, hospitals and universities too saw urban renewal as a means of supporting their interests.

However, Title I rarely met the expectations of all of its proponents, particularly supporters of public housing. In an attempt to preserve the stock of affordable housing, Title I required that each unit demolished under the banner of slum clearance must be replaced by a new unit. It also made participation voluntary, meaning suburban jurisdictions declined to apply for public housing funds. In St. Louis, for example, only 150 public housing units were developed outside the city proper through 1970 while high-rise projects like Pruitt–Igoe and Cochran Gardens came to define public housing within the city limits (Gordon 2008) (Box 3.1). As a result, public housing exacerbated segregation and further contributed to white flight.

Indeed, the most destructive aspect of urban renewal was its impacts on the lower-income, predominantly African American residents who lived in central city neighborhoods. Many people were unable to move out with upper- and middle-class whites to take advantage of suburban job and homeownership opportunities, due to racism in lending and credit markets as well as racially restrictive covenants and zoning laws. Concomitantly, rural southern African Americans migrated

---

BOX 3.1 PRUITT–IGOE

Pruitt–Igoe is one of the most legendary high-rise public housing projects of the urban renewal era (Figure 3.1). When it opened in 1954, the St. Louis housing complex was hailed as reflecting the best principles of modern architecture. Residents reported a high level of attachment and sense of community. They saw the complex as a drastic improvement over their prior housing conditions. In the words of one resident, "It was a very beautiful place, like a big hotel resort" (Friedrichs 2011). Unfortunately, after only a few years, Pruitt–Igoe endured a series of problems and came to symbolize the flaws of public housing as it was conceived under the 1949 Housing Act.

Although the initial proposal called for blending high-rise, mid-rise, and walk-up buildings, federal funding limits resulted in a final design of 33 uniform 11-story apartment buildings containing 2,870 apartments. The funding shortfall also resulted in the use of lower-quality construction materials than initially intended. In conjunction, federal funding did not provide for regular maintenance, but intended repairs to come from tenant rents, which were not sufficient to cover all costs. This set off a downward spiral. As more maintenance problems occurred, conditions worsened and people moved out. The result was even less money to address the elevator breakdowns, poor ventilation, broken windows, and lack of lighting in common spaces. Vandalism, violent crime, and drug dealing became serious problems that caused more residents to leave. By the end of the 1960s, Pruitt–Igoe was nearly abandoned and was one of the most dangerous places in the city. The entire 57-acre complex was razed in 1972.

This story is not isolated to St. Louis but was repeated in Baltimore, Chicago, Detroit, Newark, New York, and Philadelphia. Some argued that the failure of high-rise public housing and urban renewal was due to the flaws of modernist design and confirmed the important influence of environment on behavior. However, beyond design, the common denominator in each of the high-rise public housing projects was the fact that each concentrated people in residential developments without adequate access to neighborhood services and employment. As the new housing complexes welcomed their first tenants in the 1950s and 1960s, the surrounding cities were losing population and businesses were closing. Meanwhile, postwar suburbanization was in full swing, pulling jobs, people, and industry away.

Friedrichs (2011)

Figure 3.1 Pruitt–Igoe housing development
Photograph courtesy of Missouri History Museum, St. Louis

to metros in search of employment, creating a larger and poorer "black belt" in many US cities in the Northeast and Midwest (Hyra 2012). This exacerbated the race- and class-based segregation already made worse by the suburbanization process.

Additionally, although the Housing Act of 1949 initially required projects to be "predominantly residential," the vague language of the legislation and the considerable leeway given to local redevelopment agencies meant that cities could build upscale apartments, office towers, convention centers, and cultural facilities (Gotham 2001, Teaford 2000). As a result, although some early Title I projects such as those in Philadelphia and Cleveland provided mixed income, racially integrated housing, this was rare, and later revisions to the Housing Act in 1954 allowed larger proportions of development for commercial use (Teaford 2000). Boston, Denver, Los Angeles, Pittsburgh, St. Louis, and San Francisco all used urban renewal funds to uproot minority neighborhoods and build new hotel and convention complexes, upscale housing, or both.

While Title I required that families displaced by redevelopment be rehoused, the rule was rarely enforced (Fogelson 2001, O'Connor 2013). This led to significant displacement, often with little attempt by local redevelopment agencies to rehouse existing residents. By some estimates, urban renewal projects demolished 2,500 neighborhoods and 400,000 residential units occupied primarily by black households (Gotham 2001, Hyra 2012). The devastation wrought on thousands of neighborhoods resulted in the critique that urban renewal was a "downtown preservation and minority containment revitalization strategy" (Hyra 2012: 502) or, in the famous words of James Baldwin, urban renewal was "Negro removal" (Bagwell and Walker 2004).

In addition, redevelopment agencies often proposed and built enclosed commercial projects that encompassed multiple city blocks and created bubbles of opulence in seas of poverty. Many argue that projects were intentionally designed to screen out people deemed undesirable by redevelopment interests (Hyra 2012). For instance, the initial design concept for Yerba Buena Center Redevelopment in San Francisco was based on a "protected environment" scheme in which a phalanx of office towers would insulate the new commercial development from the surrounding homeless in "skid row" (Figure 3.2). Although policies have been modified to

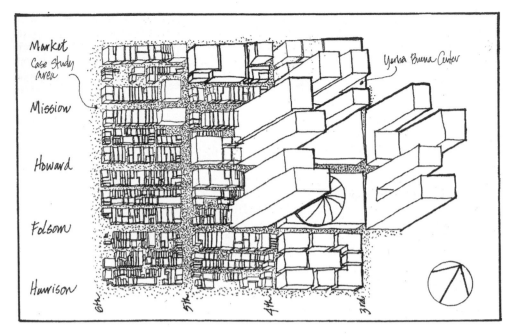

Figure 3.2   Yerba Buena Center "Protected Environment" Scheme, San Francisco, CA
Illustration by Cheryl Parker

require broader public access, higher-end commercial development continues to be designed to restrict access to lower-income users across the US (Grodach 2010).

Further, partly because agency members were appointed by local government and downtown elites, they typically focused on areas closest to the CBD with the highest potential for an increase in property values, even if these were not the most distressed areas (Fogelson 2001, Gotham 2001). Even in these cases, however, because the redevelopment process was complex, long-term and, at times, in litigation over eminent domain takings, sites would be demolished for parking lots or simply remain vacant for a decade or more. For instance, the Community Redevelopment Agency of Los Angeles declared 133 acres of land in the Bunker Hill Urban Renewal Project as blighted in 1960 and captured the land through eminent domain. Although the Agency regraded the Hill, created a new street grid, and built new infrastructure for future development, nearly two-thirds of the land remained vacant through the mid-1970s (Friedman 1978). Projects stayed vacant for so long that they earned local nicknames like "Hiroshima Flats" and "Ragweed Acres" (Teaford 2000) (Figure 3.3).

Another issue was that project planning rarely involved official community input. Nonetheless, residents in many cities still made their voices heard. In Boston, New York, Portland, Philadelphia, and San Francisco, residents protested redevelopment agency plans for demotion of their neighborhoods and often won concessions that altered the redevelopment plans (Abbott 1981, Gans 1962, Hartman 1966, Teaford 1990). The passing of the Federal Highway Act of 1956 that stimulated development of the federal interstate system led to further neighborhood demolitions and resulted in more neighborhood activism. These social movements contributed to a nation that significantly lost faith in large-scale urban projects.

Federal urban renewal has a poor record. Most urban renewal projects not only failed to improve blighted neighborhoods or improve housing quality for low-income residents, but with few exceptions such as New York's Lincoln Center and Philadelphia's Society Hill, they also did not catalyze an urban renaissance that boosted downtown property values and attracted suburbanites in great numbers as intended. Rather, the bulk of postwar redevelopment was the

Figure 3.3   "Ragweed Acres," Detroit circa 1950s
Photograph courtesy of Reuther Library, Wayne State University

result of public–private partnerships at the state and local levels (Teaford 2000). Some cities including Dallas, Houston, and New Orleans did not participate in Title I, yet here too, coalitions of government and downtown interests followed the same model of insulated, exclusive redevelopment and rebuilt downtowns that look and function remarkably similar to those that received Title I funding.

   While the urban renewal era has left a seemingly indelible mark on today's cities, some are beginning to rethink them. In Los Angeles, the City has opened the new 12 acre Grand Park to provide a public heart to the Bunker Hill redevelopment area, while in New York, Lincoln Center recently completed a $1.2 billion renovation to the project's inward-focused design. While these design interventions may improve the local environments, they do not erase the legacy of urban renewal or its impacts on urban communities. The urban renewal era nevertheless remains a high-profile example of the costs of large-scale, top-down redevelopment in which government policy is employed to generate economic growth in targeted areas and to attack the consequences of poverty rather than its causes.

URBAN REVITALIZATION AND THE WAR ON POVERTY

Although the urban renewal program continued until 1974, by the 1960s it was clear that the program was not having a significant impact on central city revitalization. The program was increasingly under fire from both left and right, as well as the impacted communities. At the same time, the Civil Rights Movement had gained momentum, demanding the elimination of segregated public services and public spaces and an end to discrimination in public education, employment, voting, and hiring practices. Although the landmark Civil Rights Act of 1964 legally outlawed unequal treatment in these areas, it was slow to have influence. Many African

Americans lived in segregated central city neighborhoods with deteriorating housing, racial prejudice, and poor job prospects because manufacturing jobs had moved to the suburbs. Throughout the 1960s, these conditions fueled civil unrest in cities of all sizes and regions including Atlanta, Chicago, Cleveland, Detroit, Los Angeles, Nashville, Philadelphia, New York, Omaha, and Rochester. The many instances of urban residents taking to the streets facing the police and National Guard, as well as the burning buildings and destroyed property, brought their desperate circumstances to the fore.

When the War on Poverty began in 1964, official poverty rates had climbed to 19% nationally. The rates in predominantly African American urban neighborhoods were much higher (Sherman, Parrott, and Trisi 2014). While urban renewal helped city centers build new corporate headquarters and office space, it had done little to address the growing joblessness, poverty, and deteriorating housing conditions. Instead, many charged that urban renewal and other federal policies contributed to the disadvantage that urban residents experienced since federally supported suburbanization of jobs and investment sharply contrasted with the continued decline of many urban areas. All these factors—the Civil Rights Movement, urban unrest, and the enduring geography of inequality—helped to propel federal urban policy in a new direction.

In 1964, President Johnson declared an "unconditional war on poverty in America." In his first State of the Union address, he proclaimed that:

> This administration today, here and now, declares unconditional war on poverty in America. I urge this Congress and all Americans to join with me in that effort … Our chief weapons in a more pinpointed attack will be better schools, and better health, and better homes, and better training, and better job opportunities to help more Americans, especially young Americans, escape from squalor and misery and unemployment rolls where other citizens help to carry them.

> (Johnson 1964)

With the War on Poverty, the Johnson administration initiated an important shift in the federal approach toward urban redevelopment. Officially designated the Economic Opportunity Act of 1964 (EOA), the War on Poverty consisted of a broad set of programs meant to establish an institutional base to tackle poverty. It included significant components intended to rectify the race-based exclusion from the country's postwar growth and the uneven development that defined US cities. Many important EOA programs focused specifically on community development. This included children- and youth-centered programs such as Head Start, JobCorps, and Neighborhood Youth Corps; the VISTA volunteer program; small business assistance and loans; and the formation of the Department of Housing and Urban Development (HUD). The EOA also led to programs meant to address the flaws of urban renewal including the Community Action Program, Special Impacts Program, and Model Cities.

Key to urban revitalization under the War on Poverty was a shift from the place-based, bricks and mortar approach under urban renewal to an emphasis on comprehensive, people-based programs that attempted to foster human capital or the skills and abilities that enable individuals to take advantage of available opportunities. People-based programs focus efforts directly on individuals or households through education, workforce training, or direct subsidies, or access to food, housing, health care, or transportation. Instead of the large-scale rebuilding projects that characterized urban renewal, EOA programs turned toward neighborhood economic development and community-led social programs alongside rebuilding efforts. Programs placed considerable emphasis on community participation in the planning processes and, at least initially, funded community organizations directly rather than funneling money through city hall and redevelopment agencies.

The Community Action Program (CAP) was the first urban program under the War on Poverty to take this comprehensive community-based approach and was intended as a direct response to many of the problems with urban renewal. The program established Community Action Agencies

(CAA), which were intended to coordinate federal and neighborhood revitalization and poverty programs and to distribute federal funds for a variety of efforts including job training, education, and youth programs. The CAAs were community-based organizations (CBOs) comprised of and organized by residents from poor communities outside the official channels of local government. Because local determination and the decentralization of power were central objectives of the CAP, it required that CAAs "ensure the 'maximum feasible participation' of the poor" across a range of areas from housing to unemployment and education (Clark 2000, O'Connor 2013). In this way the active engagement of residents in CAAs would not only improve neighborhood conditions directly, but also build the capacity of poor communities to act on their own behalf into the future.

Local governments, however, both protested the bypass of federal funds directly to communities and feared that CAAs could mobilize African Americans in high-poverty neighborhoods into political movements (Clark 2000). Just one year after the establishment of CAP, the 1965 Conference of Mayors threatened to pass a resolution against the program. In addition, local antipoverty organizations had difficulties coordinating, which added fuel to the fire to eliminate CAP. As a result, the program quickly saw its support dwindle at the federal level. While CAAs are still in operation today, most are no longer in urban areas and operate well-below the capacity initially intended.

With the future of CAP in question, the EOA was amended in 1966 to incorporate the Special Impacts Program (SIP). SIP emerged as a successor program that attempted to balance the objectives of CAP while responding to the political pressures from local governments. The program established CBOs that could receive block grants to design, finance, and administer their own revitalization programs in high-poverty neighborhoods. Programs often took place in partnership with the private sector and focused on bringing economic opportunity to impoverished neighborhoods typically through business development and job training programs. Some CBOs received criticism for setting up separate structures based on black-run community operations and white-led private investment and finance and, like other EOA programs, SIP showed few quantifiable results (O'Connor 2013). However, when the EOA expired in 1981, the program had funded 35 CBOs, 32 of which were still active in 2000 (Clark 2000).

Perhaps the best-known revitalization effort during the War on Poverty era is the Model Cities Program. Also established in 1966, the impetus behind Model Cities was recognition that existing programs were too small and dispersed to adequately address concentrated poverty and ameliorate many urban problems. The initial Model Cities concept called for a demonstration program in which the federal government would provide concentrated funding and technical assistance for comprehensive revitalization efforts in a limited number of cities, which could serve as models for other cities to emulate (Haar 1975). The program was administered by the newly formed HUD and managed by local demonstration agencies, working under the Mayor's office, which received grants to fund the construction of new housing and infrastructure alongside job training, education, health, and other social service programs.

The initial idea of demonstration projects never materialized, however. In total, 150 sites, including rural and small towns, were selected to participate in the Model Cities program. They received just half the budget initially proposed. Spread thin, the program showed little direct impact and began to be phased out in 1969 when Richard Nixon took office.

The War on Poverty was a concerted effort to address what were seen as the causes of poverty—the lack of resources, education, and skills necessary to participate in the workforce. Underlying War on Poverty initiatives was the assumption that a lack of human capital led to disadvantage. Racial discrimination led to the situation where people of color had restricted access to good jobs, education, and housing, and therefore contributed to a situation where they disproportionately had lower educational attainment and job skills. Nevertheless, it failed to address racial inequality as a serious barrier to success at the same time. Nor did programs address the suburbanization of jobs and the early signs of deindustrialization in many East Coast and Midwest cities. The framers of the policy did not acknowledge that even with equal skills, qualification, and resources, people of color faced discrimination in employment, both in hiring and promotion, as well as in housing and education (Brown 1999). Without dismantling structural racism, efforts to increase human capital were less effective in alleviating poverty than they otherwise would have been because even with more resources and education, people of color still faced discrimination when seeking work and good housing. To this end, the Civil Rights Act of 1968, often referred to as the Fair Housing Act, was one step forward. The Fair Housing Act prohibited discrimination in selling, renting, or financing housing. However, it was not enough to undo centuries of racist practices.

NEW FEDERALISM AND THE ROOTS OF NEOLIBERAL URBAN POLICY

The 1970s ushered in a major shift in federal urban policy and the thinking driving policymaking at all levels. Over the decade, most War on Poverty programs were dismantled or defunded, urban renewal officially came to an end, and cities began to endure more economic hardship as the national economy failed to maintain the high levels of postwar growth. The Nixon and Ford administrations took federal urban policy in a radically different direction by attempting to reduce federal influence and shift decision-making to states and cities.

Their political agenda, dubbed the New Federalism, was rooted in laissez-faire governance. The policy shift reduced restrictions on how localities could use federal funds, minimized federal review and monitoring, and encouraged the application of market-based programs. Spending on urban programs increased slightly under the Nixon and Ford administrations. However, with greater autonomy, the money was spread over a wider range of initiatives. The neighborhoods and residents in greatest need of assistance tended to fare worse. These policy maneuvers helped to birth neoliberal urban policy, discussed in chapter 4. Like the New Federalism, neoliberal governance emphasizes individual initiative over community and social benefit policies, and seeks to develop programs that strengthen market opportunity.

The Housing and Community Development Act of 1974 defined the policy approach of this regime. First, the Act ended the urban renewal program and shifted public housing policy from new construction to an emphasis on individual household rent subsidies under the Section 8 program, or Housing Choice Vouchers as it is known today. In brief, rather than concentrating on the construction of public affordable housing, this program provides individual residents who need housing assistance with vouchers that cover a portion of their rent. In effect, Section 8 furthered the objectives of the New Federalism by moving housing policy out of the public sector and into the private market. We discuss the function and ramifications of these policies in more depth in chapter 7.

The second defining funding instrument under the New Federalism was the Community Development Block Grant (CDBG) program. Under CDBG funding, existing urban programs were either rolled together or disbanded altogether in favor of allocating a block of annual funding to states and cities. This approach gave municipalities broad discretion in deciding where and how they used federal funds. Because the CDBG program encouraged cities to use the money as a catalyst for private investment, a mindset shared by local governments and business interests, comparatively well-off places including CBDs and suburban shopping centers had similar chances of funding as places experiencing disinvestment.

Toward the end of the 1970s, the Carter administration attempted to address the distribution issues of federal urban revitalization policy and redirect federal funds to places in need (O'Connor 2001). Carter, however, did not raise the funding levels for urban programs. He also continued to support private sector initiatives through the Urban Development Action Grant (UDAG) program. Similar to urban renewal, the grant provided funds for physical development as a catalyst for generating private investment and new employment opportunities. As a downtown redevelopment strategy, UDAG was quite successful. According to Frieden and Sagalyn (1989), UDAG funds helped build half of all downtown malls constructed between 1978 and 1985.

By the end of the 1970s, urban policy had turned away from the comprehensive and redistributive policies of the War on Poverty era and toward a path that emphasized more piecemeal, localized efforts to generate private investments. Although the hope was that those benefits would trickle down the economic ladder and lift up poor communities, the outcomes were uneven at best. As we will see in chapter 4, the New Federalism opened the door for further experiments in the market-driven approach that defines neoliberal governance, with major implications for urban revitalization.

SUMMARY

This chapter focused on the key federal urban revitalization policies enacted between the 1940s and the 1970s. We examined three distinct policy phases: urban renewal, the War on Poverty, and the New Federalism. Despite their differences, each phase maintained an emphasis on place-based revitalization efforts and, with the exception of some War on Poverty programs, tended to prioritize the production of new built environments and took on a decidedly market-driven emphasis. Further, despite the wide variety of programs and approaches, federal urban policy made little effort to challenge the structural conditions that produce poverty and economic marginalization, nor did programs focus concretely on the relationship between suburban growth and urban decline. Federal urban revitalization policy did help to establish and provide a model for the growth of CBDs and helped initiate the growth of the community development movement beginning in the 1960s.

STUDY QUESTIONS

1   What are the primary differences between the three federal urban policy phases discussed in this chapter?
2   What are the shared characteristics that define federal urban policy? Why do you think these similarities exist despite the differences in policy ideology and approach?
3   What has been the legacy of urban renewal? What can we learn from this experience?
4   What are the strengths and weaknesses of the place-based approach to urban revitalization?

NOTE

1   For a more detailed history of urban and housing policy prior to the 1940s see Goetz (2013), Jackson (1986), and Radford (1996).

REFERENCES

Abbott, Carl. 1981. *The new urban America: growth and politics in Sunbelt cities*. Chapel Hill, NC: University of North Carolina Press.
Bagwell, Orlando and W Noland Walker. 2004. *Citizen King*, edited by W Noland Walker. A Roja Productions film for American Experience, WGBH Educational Foundation.
Brown, Michael K. 1999. *Race, money, and the American welfare state*. Ithaca, NY: Cornell University Press.
Clark, Robert Francis. 2000. *Maximum feasible success: a history of the Community Action Program*. Washington, DC: National Association of Community Action Agencies.
Flanagan, Richard M. 1997. "The Housing Act of 1954: the sea change in national urban policy." *Urban Affairs Review* 33(2): 265–286. doi: 10.1177/107808749703300207
Fogelson, Robert M. 2001. *Downtown: its rise and fall, 1880–1950*. New Haven, CT: Yale University Press.
Frieden, Bernard J and Lynne B Sagalyn. 1989. *Downtown, inc.: how America builds cities*. Cambridge, MA: MIT Press.

Friedman, JH. 1978. "The political economy of urban renewal: changes in land ownership in Bunker Hill, Los Angeles." Master's thesis, Department of Urban Planning, University of California Los Angeles.

Friedrichs, Chad. 2011. *Pruitt–Igoe myth: An urban history*. Unicorn Stencil Documentary Films.

Gans, Herbert J. 1962. *The urban villagers: group and class in the life of Italian-Americans*. New York: Free Press of Glencoe.

Goetz, Edward G. 2013. *New Deal ruins: race, economic justice, and public housing policy*. Ithaca, NY: Cornell University Press.

Gordon, Collin. 2008. *Mapping decline: St. Louis and the fate of the American city*. Philadelphia, PA: University of Pennsylvania Press.

Gotham, Kevin Fox. 2001. "Urban redevelopment, past and present." In *Critical perspectives on urban redevelopment*, edited by Kevin Fox Gotham, 1–31. New York: Elsevier Press.

Grodach, Carl. 2010. "Beyond Bilbao: rethinking flagship cultural development and planning in three California cities." *Journal of Planning Education and Research* 29(3): 353–366. doi: 10.1177/0739456X09354452

Haar, Charles M. 1975. *Between the idea and the reality: a study in the origin, fate, and legacy of the model cities program*. Boston, MA: Little, Brown & Company.

Hartman, Chester. 1966. "The housing of relocated families." In *Urban renewal: the record and the controversy*, edited by James Q Wilson, 293–335. Cambridge, MA: MIT Press.

Harvey, David. 1985. *The urbanization of capital, studies in the history and theory of capitalist urbanization*. Baltimore, MD: Johns Hopkins University Press.

Hyra, Derek S. 2012. "Conceptualizing the new urban renewal comparing the past to the present." *Urban Affairs Review* 48(4): 498–527.

Jackson, Kenneth T. 1986. *Crabgrass frontier: the suburbanization of the United States*. New York: Oxford University Press.

Johnson, Lyndon. 1964. "Annual Message to the Congress on the State of the Union." Speech, January 8. The American Presidency Project. www.presidency.ucsb.edu/ws/?pid=26787

O'Connor, Alice. 2001. *Poverty knowledge: social science, social policy, and the poor in twentieth-century U.S. history.* Princeton, NJ: Princeton University Press.

O'Connor, Alice. 2013. "Swimming against the tide: a brief history of federal policy in poor communities." In *The community development reader*, edited by James DeFilippis and Susan Saegert. Abingdon, UK and New York: Routledge.

Radford, Gail. 1996. *Modern housing for America: policy struggles in the New Deal era*. Chicago, MI: University of Chicago Press.

Sherman, Arloc, Sharon Parrott, and Danilo Trisi. 2014. *Chart Book: The War on Poverty at 50*. Washington, DC: Center on Budget and Policy Priorities.

Teaford, Jon C. 1986. "The twentieth-century American City: problem." *Promise and Reality (Baltimore, Md., 1986)*: 105–106.

Teaford, Jon C. 1990. *The rough road to renaissance: urban revitalization in America, 1940–1985*. Baltimore, MD: Johns Hopkins University Press.

Teaford, Jon C. 2000. "Urban renewal and its aftermath." *Housing Policy Debate* 11(2): 443–465.

US Senate. 1949. *Summary of provisions of the National Housing Act of 1949*. Washington, DC: Government Printing Office.

4
Urban restructuring, neoliberalism, and the changing landscape of urban revitalization

By the 1970s, the vision and practice of urban revitalization shifted from the focus on social welfare and direct intervention in disadvantaged neighborhoods under the War on Poverty toward a more market-oriented agenda geared toward generating economic growth. Urban policy at both the federal and local levels became focused on competing in an unstable and globally integrated economy. The priorities of local governments changed from regulating development and providing public services to seeking out new business ventures and engaging in private sector partnerships in real estate development.

Scholars frequently refer to this form of governance as neoliberalism. Proponents of neoliberalism contend that "open, competitive and unregulated markets, liberated from all forms of state interference, represent the optimal mechanism for economic development" (Brenner and Theodore 2002: 2). Neoliberalism was born out of the New Federalism of the 1970s and crystalized into the guiding force that dismantled the social welfare policies of the War on Poverty and established a variety of new, place-based revitalization efforts. In tandem, neoliberal governance served as the political framework to address the global economic crisis of the 1970s and guide the subsequent process of economic restructuring.

In this chapter, we first discuss the economic and political context out of which neoliberalism emerged to become a powerful force in reshaping urban economies and redefining revitalization policy. Next, we turn to a discussion of the permutations of neoliberal urban policy, first as an approach to managing economic crisis, and later as the apparatus to oversee economic expansion. Within this context, we discuss the key features of neoliberal urban policy with its emphasis on urban competitiveness, privatization, and incentivizing growth. We go on to show how this political philosophy bred key federal-level urban policies such as Empowerment Zones and HOPE VI, and discuss associated policy dilemmas around community revitalization and gentrification.

THE FORDIST CITY

During the post-World War II period, the US economy experienced unprecedented growth. From about 1945 until the early 1970s, large US manufacturing industries from steel and petrochemicals to automobiles and consumer electronics surged alongside a dramatic expansion of infrastructure for the distribution, handling, and warehousing of both materials and finished products. Public investment played an important role in laying the groundwork for the long period of economic growth. Billions of dollars in state and federal funds poured into new highway construction including the Interstate Highway System. Alongside the transportation network expansion, the federal government made massive investments in private military contracts and actively oversaw the creation of international trade and monetary policy (including the establishment of the US dollar as the global benchmark currency under the Bretton Woods agreement).

The mass production and consumption that defined the postwar era describes the economic and political system known as Fordism.[1] Symbolically associated with the standardized, vertically integrated mass production of the Model T, Fordism refers to economies defined by large firms that produce standardized products and pay relatively high wages that enable the workforce to afford to purchase the commodities they make. Vertical integration means that an individual developer owns and operates all of the components required to build a product, from material suppliers to sales and marketing. Fordism is also characterized by heavy public investment and centralized planning to ensure the continued growth and expansion of the national economy. Investments in urban renewal, the Interstate Highway System, and postwar defense spending all played a role in shaping the geography of urban development.

The growth of manufacturing industries supported by major state investment led to unprecedented growth for many US cities but, as we have already seen, much of this growth did not occur in the urban core. Instead, the postwar Fordist economy carved out a distinct physical environment defined by low-density, decentralized development. Across the country, from greater Detroit to Los Angeles, new manufacturing plants and office centers pulled employment to the suburban fringe. The federal tax code continued to favor new single-family homeownership and provided strong incentive for new retail development. Like their manufacturing counterparts, the large-scale builders behind suburban planned communities such as Levittown, NY and Lakewood, CA employed assembly-line production methods and vertically integrated the entire construction process. The interstate highways also subsidized residential development indirectly by connecting outlying areas, making a larger area convenient and affordable to develop. In conjunction, federal postwar defense spending boosted employment and urban growth in California and other Sunbelt states.

Rising wages and falling prices associated with Fordist mass production enabled a growing middle class to afford a new single-family suburban home. This in turn stimulated additional consumer spending as suburbanites purchased new cars and household goods. In fact, even as new home construction slowed in the 1960s, sales of cars, TVs, lawn furniture and gardening equipment, and kitchen appliances soared.

In the 1950s, few people envisioned the demise of Fordist production centers like Detroit or that other industrial powerhouses like Chicago and New York would decline and undergo major transformations to remain economically viable. Nonetheless, the postwar Fordist system began to falter in the late 1960s due to a set of contradictions internal to the economic system itself and technological changes that facilitated new patterns of investment and disinvestment.

By the late 1960s, the US economy was increasingly challenged by more nimble overseas competition and many large producers struggled to maintain the high levels of profit and productivity they generated in the postwar era. Because Fordist industries were heavily invested in large-scale mass production they could not easily adapt to changing circumstances. This, coupled with mounting municipal and federal debt, led to higher prices, inflation, and a scaling back in worker support and wages. In Rust Belt cities like Detroit and St. Louis that were heavily dependent on manufacturing, the process was more prolonged and started as early as the 1950s. Automakers and other manufacturing industries began to close and relocate plants to other parts of the country, particularly in low-wage Southern states. Manufacturing employment in Detroit declined by 31% between 1954 and 1958 (Sugrue 2014). St. Louis experienced a similar decline during this period. By 1979, its manufacturing employment was slashed in half (Gordon 2008).

By the early 1970s, the long period of growth pivoted into a national economic crisis no longer contained in particular Rust Belt cities. The mounting economic problems were hastened by the Arab oil embargo, which was put in place to protest the support of Israel by the US and other countries in the 1973 Arab–Israeli War. This increased the price of gas and an array of manufacturing inputs that relied on petrochemicals. Alongside this, a variety of changes in international monetary policy quickly led to one of the worst stock and property market crashes since the Great Depression. Manufacturing employment, the driving force of the US economy, began to collapse. While it would recover in fits and starts through the late 1970s, the era of mass production was clearly on the wane.

UNEVEN DEVELOPMENT IN THE FORDIST CITY

Despite many gains, the Fordist city was built on inequality. While hiring practices varied by firm, industry, and place, African Americans and other people of color were often denied suburban office and manufacturing employment. Although they did find new job opportunities in the growing postwar manufacturing economy, they were disproportionately restricted to unskilled and semiskilled work and faced hiring discrimination (Teaford 2006). Moreover, the suburbanization and movement of jobs out of the industrial heartland contributed to weakening employment opportunities for many African Americans who lived in industrial city centers. While overall unemployment in Detroit grew from 7.5% in 1950 to 11.7% in 1980, it nearly doubled for African Americans, leaping from 11.8% to 22.5% during this period (Sugrue 2014).

Alongside hiring and promotion discrimination, racially restrictive covenants were enacted in postwar subdivisions like Levittown, while suburban municipalities around cities from St. Louis to Los Angeles fought to keep African American homeowners and housing projects out of their boundaries (Jackson 1985). Both employment and residential exclusion therefore steered many African Americans into segregated communities away from areas where suburban development was booming. Landmark civil rights legislation in 1964 and 1968 legally prohibited discrimination in labor and housing markets and the War on Poverty programs provided a limited safety net, but they did not remedy the deep-seated inequality. As a result, there was a significant segment of the population that was left out of the economic boom, stranded in the deindustrializing central city zones and other areas left behind during this period of economic growth. In short, the combination of deindustrialization, restricted employment opportunities, white flight to the suburbs, and center city segregation stigmatized black communities and ultimately fueled racial conflict.

Struggling central city areas experienced the economic decline hardest. Already grappling with the loss of their economic base and the flight of white middle-class property owners to the suburbs, many cities faced severe fiscal challenges. With diminishing tax revenues and increasing pressure on social welfare programs in the 1970s recession, cities accrued sizable budget deficits in an attempt to cope with unemployment and the loss of industry. Most dramatically, New York City was pushed to the brink of bankruptcy in 1975. Although cities like New York appealed for aid, the federal government made its position clear, as infamously portrayed on the cover of the New York *Daily News* in 1975 (Figure 4.1). Rather than identifying the overextended economic system and 30 years of suburbanization that drew money out of so many major cities, critics pointed to bloated municipal budgets, particularly related to overspending on social programs, as the cause of the economic malaise. This opened the way for neoliberal urban policies to dismantle the social safety net built during the War on Poverty and refocus urban revitalization policy into the 1980s.

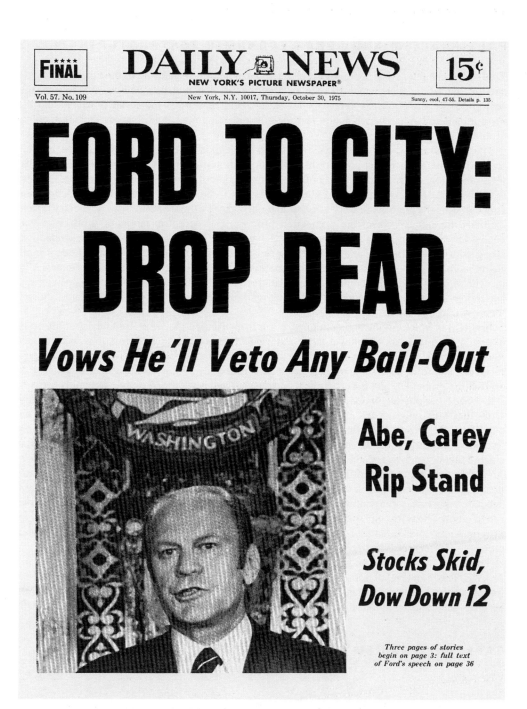

Figure 4.1   New York *Daily News* headline, October 30, 1975, "Ford to city: Drop dead"
Photograph courtesy of New York *Daily News*

POST-FORDISM AND URBAN ECONOMIC RESTRUCTURING

This set of unstable, crisis conditions brought about a major restructuring of the Fordist economy and the institutions that support it. Scholars describe this transition as post-Fordism (Amin 2011, Jessop 1995). Post-Fordism refers to the realignment of Fordist industrial practices and the institutional apparatus, which in turn reshape urban development and urban revitalization policy (Table 4.1).

The key distinction between post-Fordism and Fordism is the reliance on specialized, flexible production methods. Under post-Fordism, the production process decentralized and the vertical integration that characterized the Fordist system become disintegrated. In other words, rather than a single firm owning all of the materials and suppliers involved in production and concentrating the majority of work at a central point, a range of specialized firms or subcontractors handle different aspects of the production process. The decentralization of production was enabled by new computing and information technologies. Technological changes in production, along with subcontracting, allow firms to speed up turnover time and therefore to adapt more easily to changing demand and target niche markets. This combination of factors has enabled the growth of very large corporations dealing in sectors as different as information technology, apparel or TV, radio and film to dominate their markets but often without directly manufacturing their own products.

In this way, mass production continues, but it is distributed globally. Routine assembly work is now moved to areas with lower labor and production costs. Large corporations that produce low and mid-skill labor-intensive products including components for automobiles, apparel, footwear, and toys, have splintered the production process, moving various stages of production around the globe. US and European firms downsized and outsourced routine assembly work to low-wage locations such as China, Taiwan, Indonesia, and Mexico. As a result, already struggling cities that were once centers of auto or steel production, such as Detroit, MI, Youngstown, OH, and Gary, IN, experienced further declines in employment as production was contracted out overseas. Similarly, more diversified economies in Los Angeles and New York, for instance, saw sizable sectors such as apparel diminish.

At the same time, corporations retained managerial and other skilled labor in home country headquarters and established command centers in global cities where highly skilled labor and specialized services concentrated (Sassen 2012). This occurred because post-Fordist industries relied more heavily on specialized labor and expertise. Industrial designers, software programmers, and specialized accounting, for instance, help firms to respond to fluctuations in consumer preference, identify niche markets, and create more varied product lines.

The growth of finance and business services was also a central facet of the restructuring process. The high level of transactions and coordination necessary to manage the global dispersal of industry and trade necessitates an intricate network of finance, legal, insurance, and accounting services and telecommunications infrastructure. As Sassen (2012) documents, in the 1980s

TABLE 4.1 FORDISM AND POST-FORDISM

Fordism	Post-Fordism
Mass production	Flexible production
Vertical integration	Functional specialization
Centralized production process	Decentralized production process
Standardized products	Product differentiation
Mass consumption	Niche markets
Middle class	Bifurcated labor force

foreign direct investments (FDI) grew three times as fast as export trade. The bulk of these investments were in business services, not manufacturing and materials extraction, which defined the majority of investments in the past.

The growth in FDI and the complex financial transactions helped to create a new role for cities as "strategic sites for the management of the global economy" (Sassen 2012). Firms that depend on specialized services and telecommunications infrastructure locate where they have immediate access to these requirements. As a result, financial services and other knowledge-intensive industries in the global economy gravitate to where they can access the concentration of services and infrastructure necessary to manage their world-wide operations. Cities including New York, London, Paris, and Tokyo are leading hubs of the global economy, but because the system depends on a network of national and regional services as well, US cities including Chicago, Dallas, and San Francisco also serve as significant command and control centers, where major corporate headquarters and specialized business services concentrate.

In this shifting economic landscape, some places experienced the revitalization of traditional craft-based industries. One of the most famous examples of post-Fordism is the rise of the "Third Italy" (Piore and Sabel 1984). Diverse industries including woodworking, ceramics, furniture, and higher-end apparel employ skilled workers engaged in customized, small-batch production using flexible manufacturing technologies and work practices. In other places, industry agglomerations successfully coordinate localized networks of complementary firms governed by strong institutional support. These characteristics have been associated with the rise of high-technology centers such as Silicon Valley as well as the growing film and cultural industry clusters that began to flourish through concentrated networks of vertically disintegrated production and specialized labor and expertise (Saxenian 1996, Scott 2000).

URBAN RESTRUCTURING AND UNEVEN DEVELOPMENT

As the restructuring process has transformed local economies, it has produced a new round of uneven development both within and between cities. When manufacturing relocated to new countries around the world, former manufacturing hubs in the US lost middle-class jobs that helped to drive the postwar urban growth. Some cities experienced economic recovery and growth as hubs of advanced services, but others did not easily adapt. Cities that specialized in declining manufacturing sectors have been especially vulnerable during and after the restructuring process. The built environment and infrastructure in many Rust Belt cities was set up for a prior economic period and their workforce ill-equipped to transition to the new economy. These cities, so dependent on manufacturing economies and lacking a diverse economic base, have struggled with rising unemployment, poverty, and homelessness as well as vacant residential property and abandoned industrial districts. Since 1970, Buffalo, Cleveland, Detroit, and Pittsburgh have each lost over 40% of their population and median household incomes have dropped in each city, ranging from 10% in Pittsburgh to 35% in Detroit (Hartley 2013).

In cities well-positioned in the new economy, the rise of industries dependent on highly skilled professionals combined with the loss of middle-wage industries caused the labor market to become increasingly bifurcated. For example, the top 10% of wage earners in New York saw their income grow over 20% between 1977 and 1986, while the bottom 20% saw their income decline. Moreover, even as incomes and wages grew, poverty increased from 15% to 23% during the period (Mollenkopf and Castells 1991). The advanced services that created jobs in CBDs offered high wage opportunities for professionals with high levels of education. Between 1977 and 1987, finance and business services accounted for 70% of new jobs created in New York (Castells 1989).

As in prior decades, urban restructuring in the 1970s and 1980s revealed differential racial impacts. Minorities, particularly African Americans and Latinos, who had proportionally larger employment in manufacturing, were not prepared for the new jobs and were hit hardest by economic restructuring. As Sugrue (2014) documents in Detroit, the automotive industry was the city's largest employer of African Americans, with some plants reporting as many as 60% black employees in 1960. The skills mismatch between the new jobs in technology and advanced services and the existing skills of central city African American and Latino workers created a situation where large segments of urban populations were not equipped to take advantage of available opportunities (Massey and Denton 1987, Wilson 1997).

New employment opportunities also expanded at the bottom of the wage spectrum, primarily in basic services such as retail, food services, and other low-skilled jobs including janitorial and secretarial work to support the demands of those working in the advanced services and tech sectors. However, these jobs offer considerably lower wages and very few benefits compared with traditional manufacturing jobs. Much of this employment has been taken up by immigrants and women entering the labor force. The result is a "dual city" defined by a workforce that is highly polarized and segmented (Castells 1989, Mollenkopf and Castells 1991).

As a result, nationally, while concentrations of affluence increased, so too did concentrations of poverty. Neighborhoods defined by concentrated poverty were nearly three times more likely to be predominantly African American than white in 1970 and the rate increased over the 1980s. Concentrated affluence was most pronounced in those cities that also contained the highest concentrations of poverty including Chicago, Detroit, New York, and Newark (Massey and Denton 1993).

Figure 4.2 shows the employment trends at the national level through the restructuring period. By 2000, employment in manufacturing had diminished and was almost overtaken by advanced services (including finance and business services). During the same time period, basic services grew exponentially to overtake manufacturing as the largest employment base in the US.

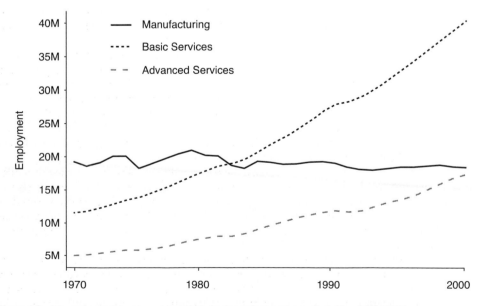

Figure 4.2   US employment in advanced services, basic services, and manufacturing, 1970–2000
Bureau of Labor, www.bls.gov/ces/data.htm

THE EMERGENCE OF NEOLIBERAL URBAN POLICY: CRISIS MANAGEMENT AND URBAN ENTREPRENEURIALISM

Global economic restructuring was not guided by the natural course of the market but by social and political forces. Neoliberalism was the overarching political philosophy that guided the transition, with vast implications for urban redevelopment and governance (Brenner and Theodore 2002, Hackworth 2007, Harvey 2005, Peck and Tickell 2002). The main tenet of neoliberalism is that human well-being can best be advanced by liberating individual entrepreneurial freedoms and skills within an institutional framework characterized by strong private property rights, free markets, and free trade. The role of the state is to create and preserve an institutional framework appropriate to such practices. But beyond these tasks, the state should not venture (Harvey 2005: 2).

In the urban context, the neoliberal doctrine promotes creating and maintaining opportunities for private sector investment that will in turn produce revenue for the city or reduce its budget outlays. Government intervention prioritizes economic growth over the redistribution of economic benefits or ensuring a social safety net, as we saw during the War on Poverty. Because government does not possess the means to make direct investments, its primary role under neoliberalism is to create the conditions for private sector investment whether in property redevelopment, new industry development, or the privatization of public services. According to neoliberal thought, this, rather than direct government intervention, produces the greatest level of social and economic benefits because the benefits from robust economic activity will trickle down from entrepreneurial elites to the general population.

Neoliberal governance gained force in the 1980s as a form of crisis management. Proponents, most notably Ronald Reagan in the US and Margaret Thatcher in the UK, exploited the Fordist crisis conditions to advance their neoliberal agendas aimed at economic recovery and retiring the social welfare programs put in place in the 1950s and 1960s. This initial stage of neoliberalism, which Peck and Tickell (2002) label "roll-back" neoliberalism, blamed the economic and fiscal crises on bloated government. Proponents used the crisis to justify the termination of important social programs, deregulate trade and industry, and privatize public services. Municipal and national governments throughout North America and Europe as well as South Africa, Australia, and China among other countries pursued this program to varying degrees to steer the restructuring of their economies (Harvey 2005).

Thatcher and Reagan went beyond Nixon's New Federalism to further roll back commitments to social welfare and urban redevelopment programs. Between 1980 and 1990, under Presidents Reagan and George Bush I, the federal minimum wage fell 30%, economic development assistance plummeted 78%, Community Development Block Grant funding was slashed in half, and Urban Development Action Grants and revenue sharing programs were eliminated (Gotham 2001, Harvey 2005, O'Connor 2013).

Between 1980 and 1992, federal government assistance to local governments dropped from 14.3% to just 5% of municipal budgets nationwide (Gotham 2001). Rather than assist local governments in coping with deindustrialization, the intent of the federal cost-cutting measures was to discipline cities into becoming market actors and working within the neoliberal framework. Local governments were forced to adapt to the fiscal austerity measures and make up the funding deficit through limited local resources. These funding reductions had the greatest impact on the urban poor whose numbers grew from 24 million in 1978 to 33 million in 1984 (Florida and Jonas 1991).

At the same time, neoliberal policy concentrated on encouraging the privatization of public services and creating a pro-business environment defined by minimal state regulations and low taxes. To this end, substantial deregulation took place in finance, telecommunications, and other industries as well as environmental protection, occupational safety, public utilities, and transportation infrastructure (Harvey 2005, Sager 2011).

Fiscal austerity, deregulation, and privatization are the key tenets of neoliberalism, which guided the economic transition (Table 4.2). As a result of these measures, cities made a pronounced shift from managerialism to entrepreneurialism (Harvey 1990). Under a managerial system, local governments focused more on regulating development and providing urban services. With the change to an entrepreneurial regime, municipalities began to take a more competitive and speculative stance toward governance and planning. Federal neoliberal policy coerced many municipalities into cutting and privatizing municipal public services as measures to reduce the size and budget of local governments. At the same time, they made more concentrated efforts at courting private investment to realize their revitalization and development goals. In short, neoliberal urban policies amounted to guiding a new "spatial fix" under post-Fordism.

This shift had major repercussions for urban policy. The destruction of Fordist-era safety net programs and funding streams made way for the "roll-out" of market-oriented neoliberal urban policy (Peck and Tickell 2002). Municipalities made more concentrated efforts at courting private investment to realize their revitalization and development goals.

As Eisinger explains:

> The entrepreneurial state … seeks to identify market opportunities not for its own exclusive gain but on behalf of private actors whose pursuit of those opportunities may serve public ends. In its role of partner the state has become a risk-taker, a path-finder to new markets, the midwife to joint public–private efforts to develop and test untried technology … supplying the necessary wherewithal to stimulate new private-business formation … Underlying the actions of the entrepreneurial state is the assumption that growth comes from exploiting new or expanding markets. The state role is to identify, evaluate, anticipate and even help to develop and create these markets for private producers to exploit, aided if necessary by government as subsidizer or coinvestor.
>
> (Eisinger 1988: 9)

Localities took an entrepreneurial stance in response to the geographic and sectoral shifts in economic development that resulted in the outsourcing of traditional manufacturing industries and the rise of advanced services and consumption industries. In this context, local governments rebuilt their economies by wooing growth industries and redeveloping older commercial and industrial spaces for tourism and consumption.

TABLE 4.2 NEOLIBERAL GOVERNANCE AND THE TRANSITION FROM FORDISM TO POST-FORDISM

	Fordism	*Post-Fordism*
Governance	Managerial welfare role	Entrepreneurial role
	State funding for municipal government activities	Fiscal austerity
	Public service and infrastructure provision	Deregulation and privatization of public services
	Regulate development	Facilitate development
Urbanization	Suburbanization	Gentrification
	Comprehensive planning	Catalytic projects
	Large-scale development	Themed development
	Working-class city image	Marketing business climate and consumer amenities

A key focus is on place marketing. While urban boosters have long attempted to promote a positive image of their city, place marketing has become more widespread and professionalized since the 1980s. As cities assumed a more competitive stance, they focused on promoting a friendly business climate, a cosmopolitan image, and an attractive quality of life. Some cities have become famous for their marketing slogans—think of I ♥ NY, or Austin "The Live Music Capital of the World." Through promotional material and slogans, cities attempt to hide their industrial past and selectively depict their history and culture, or to demonstrate that they possess a new, dynamic business environment. However, 1980s slogans such as Detroit "Renaissance City," "Cleveland Rocks," or "Newark, on a Roll" were not always so successful. Although the level and extent of neoliberal-dominated urban policy has varied from place to place, the competitive stance and emphasis on "selling places" has become common in most localities.

Another hallmark feature of the entrepreneurial approach is the focus on creating private sector demand for a locality. Facing increasing interurban competition, municipalities adopt a range of strategies to induce private sector investment. This often consists of strategies to reduce the cost of doing business through various forms of tax abatement, low-interest loans, and land grants. Planning agencies streamline permitting processes and allow for flexible zoning to encourage development in targeted areas. They focus on developing large, catalytic projects rather than coordinating development through comprehensive planning.

As a result, cities devote more attention to the development of iconic architecture, flagship cultural institutions, central public spaces, and major events like the Olympics (Evans 2003, Grodach 2009, Kearns and Philo 1993). Baltimore's Inner Harbor, Faneuil Hall in Boston, Navy Pier in Chicago, and Battery Park City in New York are all early examples of major redevelopment projects meant to attract tourists and provide amenities for a growing white-collar workforce while improving the city image.

Most of these redevelopment projects were realized through public–private partnerships, one of the "foundations of neoliberal governance" (Hackworth 2007: 61). Public–private partnerships typically position the municipality as facilitator in that they subsidize and streamline the development process, though they also may become co-developers in a project. Public entities stake strategically located public land and offer attractive grants and loans as incentives for developing key property and extracting benefits for development rights.

Proponents claim that public–private partnerships help cities to compete and generate interest in hard-to-develop property (Sagalyn 2007). Critics argue that the emphasis on developing property for entertainment, consumption, and office uses amounts to a subsidy for the wealthy and encourages gentrification in the urban core (Smith 2002, Zukin 1989). Further, like urban renewal projects before them, the joint public–private ventures rarely incorporate adequate community participation and they target those places with the strongest capacity for increasing the tax base and property values. As a result, this strategy ignores those places that have fewer locational advantages and development potential, but have endured decades of underinvestment. This in effect creates a bubble of affluent central city investment while leaving the rest of the central city to deal with job loss, crime, and abandonment (Hackworth 2007, Squires 1989, Harvey 1990).

RESHAPING NEOLIBERAL URBAN POLICY: COMMUNITY REVITALIZATION OR GENTRIFICATION STRATEGY?

Into the 1990s, state and federal urban policy showed signs of tempering the market-oriented neoliberalism of the prior decade by incorporating social and community considerations into revitalization programs. While 1980s-era urban policies helped to redevelop and increase the value of select central city properties, alleviating the poverty and unemployment that came in the wake of urban restructuring was often not a priority. In fact, given the focus on property appreciation and the development of up-market tourist bubbles, redevelopment often contributed to social and economic polarization. Urban policy in the 1990s did not abandon the market-driven emphasis, but it attempted to reformulate neoliberal urbanism to address these issues.

One of the most high-profile proponents of market-driven urban revitalization in the 1990s and 2000s was Michael Porter and his Initiative for a Competitive Inner City (ICIC). Porter argues that the route to central city revitalization is essentially through promoting the "competitive advantages" of poverty-stricken neighborhoods. This includes the high unmet demand due to the dearth of retail and grocery stores in struggling neighborhoods, and strong resident purchasing power due to higher-density neighborhoods. He also notes that businesses can take advantage of a captive labor market (i.e. high unemployment) and a strategic central location close to the CBD typically with good highway access (Porter 1997). According to an ICIC report, inner-city neighborhoods in the 100 largest US cities possess $120 billion in annual retail spending, yet one-third of that spending occurs elsewhere (Coyle 2007). In short, Porter's route to revitalization showed businesses that they could profit from disadvantaged communities where demand was not being met. Not only will poor communities see new investment, but corporate interests will turn a profit. The promotion of the untapped advantages of investing in retail-starved central city neighborhoods, "the next big retail frontier" according to *Commercial Investment Real Estate* magazine, has encouraged a variety of corporate and big-box retailers such as Target and Walmart to downsize and open stores there (Drummond 2007) (Figure 4.3). Sometimes they open in "food deserts," neighborhoods with few grocery options—and little competition.

While Porter's profit-oriented argument has helped to bring much-needed investment to many central city neighborhoods, it fails to address the fact that business attraction alone rarely results in true urban revitalization. Many new businesses bring in skilled labor from outside the community or they provide low-wage service sector employment. In addition, business development alone cannot solve the range of social issues that need to be addressed in struggling communities. Porter does stress the importance of job training and building entrepreneurial skills for residents in low-income neighborhoods, but this has been a greater challenge and one that

Figure 4.3   Walmart, H Street NW, Washington, DC
Photograph by Ed McMahon

stems from larger structural issues related to urban development and racial discrimination. While Porter offers some practical advice for addressing economic opportunity in distressed neighborhoods, he is silent on tackling the interconnected social issues that work against economic development there.

The move to a more socially considerate neoliberalism is further epitomized by the "Third Way" policies of Bill Clinton in the US and Tony Blair in the UK. The Third Way attempts to integrate economic and social imperatives through policy (Giddens 2013). To be sure, this is not a return to the social redistribution and welfare emphasis of the 1960s. Rather, established welfare programs were transformed into "workfare" programs that focused on getting people back to work quickly and imposed mandatory requirements on recipients of public benefits.

In urban policy, Third Way proponents sought to merge social and community concerns with market-led revitalization policies that target and incentivize place-based property and business development. Third Way policies soften the emphasis on austerity of the Reagan–Thatcher era and incorporate community involvement and workforce participation. They continue to emphasize partnerships, but look to nonprofit philanthropic foundations and community and faith-based organizations as well as the private sector to carry out urban revitalization initiatives. While some assert that Third Way policies provide new opportunities and a voice for marginalized communities, others argue that the move toward community collaboration and employment are simply maneuvers to further downsize government and outsource publicly provided services (Peck and Tickell 2002).

The Empowerment Zones/Enterprise Communities initiative exemplifies this shifting face of neoliberal urban policy. Initiated in 1994 under President Clinton, the program blends an emphasis on entrepreneurialism and competition with community engagement. EZ/ECs are intended to spur job creation and economic growth in distressed communities primarily through tax incentives and the removal of government regulations. The Enterprise Zone concept was first introduced in the UK in the late 1970s and was embraced in the US by the Reagan administration, but implemented at the state level only. By 1983, nearly half of all US states established an Enterprise Zone program that offered varying levels and types of tax incentives, special capital financing, and a reduced regulatory environment.

The federal EZ/EC program provides federal block grants to support business development and local hiring in designated areas (McCarthy 1998, Oakley and Tsao 2006). As we explain in more detail in chapter 7, cities compete for awards, which they may use to offer tax credits to incentivize business investment and local hiring in targeted low-income communities. The EZ initiative also reduces regulatory requirements for businesses and includes a community participation component in the planning process. Enterprise Zone initiatives exhibit mixed outcomes in terms of both economic development and community participation.

Alongside incentivizing business development in targeted areas, a key focus of the neoliberal urban policy shift has been an attempt to deconcentrate poverty and attract middle-class households to central city areas. This ongoing emphasis stems from a belief that high levels of segregation and concentrated poverty isolate and compound problems such as crime and joblessness (Massey and Denton 1993, Wilson 1997). Additionally, with the growth of advanced services in many CBDs, local governments recognized and attempted to capitalize on the upper-income workforce demand for housing in surrounding low-income neighborhoods (Fainstein 2001, Hackworth 2007).

In 1992, the Clinton administration initiated Housing Opportunities for People Everywhere (HOPE VI) in response to these issues. The intent of the housing program was to break up high-poverty neighborhoods by demolishing older, distressed public housing complexes and replace them with mixed-income developments. The goal of HOPE VI, which was funded through 2009, was to provide better housing for poor residents while catalyzing private investment in surrounding communities. Alongside this, residents of the demolished public housing developments received Housing Choice Vouchers (Section 8), which were intended to facilitate their move out of distressed center city neighborhoods and to moderate-income suburban areas.

As we discuss in chapter 7, HOPE VI has been a controversial program because, while it removed many aging housing projects, it also displaced low-income residents from high-value center city property, often without providing adequate replacement housing or necessarily improving their lives or opportunities (Goetz 2013, Popkin et al. 2004).

The federal government also worked to make central city property more attractive for development through the 1993 Brownfield Initiative. This Environmental Protection Agency program funded the acquisition and remediation of many contaminated former industrial sites that blighted central city communities. At the same time, the Brownfield Initiative pursues a market-led approach based on "industrial triage." This strategy, which we explore in more detail in chapter 10, calls for remediating those sites with the highest development potential. As a result, many neighborhoods with abandoned industrial property are less likely to receive assistance (Fitzgerald and Green Leigh 2002).

THE GREAT RECESSION, COMPREHENSIVE COMMUNITY INITIATIVES, AND THE SHIFTING GEOGRAPHY OF POVERTY

Behind Third Way urban policy, the ongoing integration of the global economy and the deregulation of financial markets helped to fuel a new round of urban redevelopment in the 2000s. As knowledge-intensive, advanced service sector jobs multiplied, so too did demand for central city neighborhoods and amenities. This demand, coupled with the loosening flow of credit into central city areas where property values had been suppressed, enabled substantial investment in new commercial and residential real estate projects.

The gentrification of central city neighborhoods spread beyond those in major urban centers like New York and Los Angeles to emerging global cities such as Atlanta, GA and Dallas, TX (Hackworth and Smith 2001). Portland, OR and Austin, TX became magnets for the growing "creative class" attracted by these cities' lifestyle amenities and strong technology economies (Florida 2002). Local governments across the US incentivized large-scale office and residential projects, the rehabilitation of older buildings into loft-style apartments, and the construction and expansion of new arts and entertainment venues. Downtown Atlanta and Austin, Dallas's Uptown, and Portland's Pearl District are some of the prominent examples that reflect these changes (Figure 4.4).

Redevelopment projects like these have been heralded in the popular press as revitalization success stories, yet this success has come at a significant cost. The demolition of public housing and escalating property values have displaced both lower- and middle-income residents from central neighborhoods. Large-scale redevelopment has pushed many poor and minority households into inner suburban neighborhoods. Place-based renewal has been accomplished only by shifting poverty out of the central city and rendering it almost invisible in the suburbs. In fact, during the first decade of the 2000s, suburban poverty in the US became more concentrated and grew over twice as fast there as in center cities (Kneebone and Berube 2013).

Simultaneously with the large-scale gentrification of the urban core, the 2000s saw unprecedented deregulation in mortgage lending markets. Subsequently, the flow of credit and lending expanded dramatically as financial institutions bundled loans into mortgage-backed securities and sold them to global investors. Residents in underserved minority communities in the central city and inner suburbs found they could more easily obtain a home loan, but it was often in the form of a subprime mortgage with interest rates higher than a conventional mortgage (Wyly, Atia, and Hammel 2004). Potential homebuyers also found that they could purchase homes with little down payment and variable mortgage interest rates that offered favorable terms in the short term, but quickly became unaffordable.

While this enhanced access to credit enabled new homeownership and redevelopment opportunities, by 2007 borrowers began to default in record numbers—foreclosure filings leaped 75% during that year (RealtyTrac 2008). In some neighborhoods over 500 households filed for

Figure 4.4   2nd Street District, Austin, TX
Photograph courtesy of 2nd Street District and Sabrina Bean Photography

foreclosure in 2007 alone (CNN Money 2007). In short, a decade of lax regulations in financial markets enabled overextended and sometimes unscrupulous lending and historic levels of debt to accrue. The investments lost value as property owners could not make payments, resulting in the burst of the housing bubble that, alongside a drop in investor and consumer confidence, contributed to the global recession. The effects of the housing market crash rippled through urban economies resulting in significant job loss, and curtailed property and sales tax income for cities.

The Third Way brand of neoliberal governance held steady even through the great recession under both the Bush II and Obama administrations. Despite the economic crash, it remains the guiding policy regime at the national and local levels (Peck, Theodore, and Brenner 2013) though it has been reinterpreted by different political regimes (Kantor 2013). George Bush II was antagonistic toward urban policy (making repeated attempts to end HOPE VI funding). However, Barack Obama has demonstrated renewed interest in a federal urban policy agenda, particularly during his first term. His administration's two primary programs under the *White House Neighborhood Revitalization Initiative* (2012)—Choice Neighborhoods and Promise Neighborhoods—expand the focus on comprehensive, place-based revitalization through complicated partnerships and leveraged private investment. Choice Neighborhoods, which builds on the HOPE VI program, provides funding for mixed-income housing while comprehensively planning for education, health, and other public services in the surrounding neighborhood. Similarly, the Promise Zones initiative takes a comprehensive approach toward education by incorporating an array of social services into the "cradle to college" program (Smith 2011). We examine Choice Neighborhoods in more depth in chapter 7.

Nevertheless, funding for Obama's signature programs pales in comparison with the funding meant to clean up the aftermath of the financial crisis. The 2009 Recovery and Reinvestment Act increased Community Development Block Grant (CDBG) funding by $1 billion, added $4 billion toward public housing, and enhanced funding for other workforce and social assistance programs (US Department of Housing and Urban Development n.d.). In contrast, the most recent round of funding for Choice Neighborhoods and Promise Neighborhoods totaled just under $138 million in 2014. In comparison, $45 billion in aid went to Citigroup and Bank of America, and nearly $51 billion went toward saving General Motors from bankruptcy in 2009 (Propublica 2014).

SUMMARY

The 1970s ushered in a set of changes that dramatically altered US urban economies, patterns of urban development, and the contours of local and national revitalization policy. The postwar Fordist city was defined by growth in large-scale manufacturing, middle-class jobs, and suburban homes alongside a declining urban core. In contrast, under the period of post-Fordist restructuring, cities that captured advanced business services and other industries dependent on highly skilled labor garnered major income and employment growth, while leading manufacturing centers struggled.

This uneven geography of development occurred not only between cities but within them as well. High-wage, high-growth regions are also home to some of the highest concentrations of low-skill, low-wage jobs performed largely by minorities, immigrants, and women.

We also see patterns of growth begin to change. While suburban expansion continues, many central cities have been transformed through the redevelopment of CBDs, entertainment districts, and newly gentrified neighborhoods while older, inner-suburban areas experience rising poverty.

In terms of urban policy, we have seen a pronounced movement away from the targeted social and community-oriented programs of the 1960s and toward catalytic, market-oriented redevelopment projects. Beginning in the late 1970s and early 1980s, neoliberal governance emerged as a powerful force to manage the economic crisis and guide the restructuring process. The deregulation and privatization of industry and public services has been key to the dismantling of the Fordist city. In their bid to attract new development and investment, local governments target older central city areas for reinvestment and help to incentivize the post-Fordist spatial fix concentrated on the gentrification of central city neighborhoods alongside continued exurban expansion. Despite cleaner and safer downtowns, neoliberalism has not resolved the social and economic problems under Fordism, but has shifted the geography of inequality and heightened social and economic polarization. While President Obama made some effort to refocus attention on cities and urban policy, increasing suburbanization of poverty remains in a federal "policy blindspot."

STUDY QUESTIONS

1 How does the concept of post-Fordism help to explain changes in urban economies and development patterns?
2 How has neoliberalism influenced urban revitalization policy? Do particular features of neoliberalism help or hurt the potential for urban revitalization?
3 Do you feel that public–private partnerships are good for urban revitalization programs? Explain why.
4 How is urban redevelopment policy linked to the suburbanization of poverty? What are some other potential causes?

NOTE

1 Scholars coined the term Fordism to describe the "regime of accumulation," or the overarching system that shapes not only the economy but also the social and political relations, during the postwar period (Goodwin and Painter 1996, Jessop 1995, Aglietta 2000). A central insight of regulation theory is that the characteristics

that embody an accumulation regime become unstable over time and lead to a crisis. This in turn leads to a period of economic insecurity and restructuring as market actors seek new sources of profit and build new institutions to govern the new regime.

REFERENCES

Aglietta, Michel. 2000. *A theory of capitalist regulation: the US experience*. Vol. 28. London and New York: Verso.

Amin, Ash. 2011. *Post-Fordism: a reader*. In *Studies in urban and social change*, edited by Chris Pickvance and Margit Mayer. Hoboken, NJ: John Wiley & Sons.

Brenner, Neil and Nik Theodore. 2002. "Cities and the geographies of 'actually existing neoliberalism'." *Antipode* 34(3): 349–379. doi: 10.1111/1467-8330.00246

Castells, Manuel. 1989. *The informational city: information technology, economic restructuring, and the urban-regional process*. Oxford: Blackwell.

CNN Money. 2007. "500 top foreclosure zip codes: where the most foreclosures have been filed." http://money.cnn.com/2007/06/19/real_estate/500_top_foreclosure_zip_codes/index.htm?iid=EL

Coyle, Deirdre 2007. "Realizing the inner city retail opportunity: progress and new directions, an analysis of retail markets in America's inner cities." *Economic Development Journal* 6(1): 6–14.

Drummond, Sara. 2007. "New rules for retail: saturated markets and deep-pocket competitors are prompting changes in this once by-the-book investment sector." *Commercial Investment Real Estate Magazine* May/June. www.ccim.com/cire-magazine/articles/new-rules-retail

Eisinger, PK. 1988. *The rise of the entrepreneurial state: state and local economic development policy in the United States*. Madison, WI: University of Wisconsin Press.

Evans, Graeme. 2003. "Hard-branding the cultural city—from Prado to Prada." *International Journal Of Urban and Regional Research* 27(2): 417–440. doi: 10.1111/1468-2427.00455

Fainstein, Susan S. 2001. *The city builders: property development in New York and London, 1980–2000, studies in government and public policy*. Lawrence, KS: University Press of Kansas.

Fitzgerald, Joan and Nancey Green Leigh. 2002. "Job-centered economic development: an approach for linking workforce and local economic development." In *Economic revitalization: cases and strategies for city and suburbs*, edited by Joan Fitgerald and Nancey Green Leigh. Thousand Oaks, CA: Sage.

Florida, Richard L. 2002. *The rise of the creative class and how it's transforming work, leisure and everyday life*. New York: Basic Books.

Florida, Richard and Andrew Jonas. 1991. "US urban policy: the postwar state and capitalist regulation." *Antipode* 23(4): 349–384.

Giddens, Anthony. 2013. *The third way: the renewal of social democracy*. Hoboken, NJ: John Wiley & Sons.

Goetz, Edward G. 2013. *New Deal ruins: race, economic justice, and public housing policy*. New York: Cornell University Press.

Goodwin, Mark and Joe Painter 1996. "Local governance, the crises of Fordism and the changing geographies of regulation." *Transactions of the Institute of British Geographers* 21(4): 635–648.

Gordon, Collin. 2008. *Mapping decline: St. Louis and the fate of the American city*. Philadelphia, PA: University of Pennsylvania Press.

Gotham, Kevin Fox. 2001. "Urban redevelopment, past and present." In *Critical perspectives on urban redevelopment*, edited by Kevin Fox Gotham, 1–31. New York: Elsevier Press.

Grodach, Carl. 2009. "Urban branding: an analysis of city homepage imagery." *Journal of Architectural and Planning Research* 26(3): 181.

Hackworth, Jason. 2007. *The neoliberal city: governance, ideology, and development in American urbanism*. Ithaca, NY: Cornell University Press.

Hackworth, J and N Smith. 2001. "The changing state of gentrification." *Tijdschrift voor economische en sociale geografie* 92: 464–477. doi: 10.1111/1467-9663.00172

Hartley, Daniel. 2013. "Urban decline in rust-belt cities." Federal Reserve Bank of Cleveland. www.clevelandfed.org/en/Newsroom and Events/Publications/Economic Commentary/2013/ec 201306 urban decline in rust belt cities.aspx

Harvey, David. 1990. *The condition of postmodernity: an enquiry into the origins of cultural change*. Oxford: Blackwell.

Harvey, David. 2005. *A brief history of neoliberalism*. Oxford: Oxford University Press.

Hyra, Derek S. 2012. "Conceptualizing the new urban renewal: comparing the past to the present." *Urban Affairs Review* 48(4): 498–527.

Jackson, Kenneth T. 1985. *Crabgrass frontier: the suburbanization of the United States*. Oxford: Oxford University Press.

Jessop, Bob. 1995. "The regulation approach, governance, and post-Fordism: alternative perspectives on economic and political change?" *Economy and Society* 24(3): 307–333.

Kantor, Paul. 2013. "The two faces of American urban policy." *Urban Affairs Review* 49(6): 821–850. doi: 10.1177/1078087413490396

Kearns, Gerry and Chris Philo. 1993. *Selling places: the city as cultural capital, past and present*. Oxford: Pergamon Press.

Kneebone, Elizabeth and Alan Berube. 2013. *Confronting suburban poverty in America*. Washington, DC: Brookings.

Massey, Douglas S and Nancy A Denton. 1987. "Trends in the residential segregation of Blacks, Hispanics, and Asians: 1970–1980." *American Sociological Review* 52: 802–825.

Massey, Douglas S and Nancy A. Denton. 1993. *American apartheid: segregation and the making of the underclass*. Cambridge, MA: Harvard University Press.

McCarthy, John. 1998. "US urban empowerment zones." *Land Use Policy* 15(4): 319–330.

Mollenkopf, John H and Manuel Castells. 1991. *Dual city: restructuring New York*. Washington, DC: Russell Sage Foundation.

Oakley, Deirdre and Hui-Shien Tsao. 2006. "A new way of revitalizing distressed urban communities? Assessing the impact of the federal Empowerment Zone program." *Journal of Urban Affairs* 28(5): 443–471. doi: 10.1111/j.1467-9906.2006.00309.x

O'Connor, Alice. 2013. "Swimming against the tide: a brief history of federal policy in poor communities." In *The community development reader*, edited by James DeFilippis and Susan Saegert. Abingdon, UK and New York: Routledge.

Peck, Jamie and Adam Tickell. 2002. "Neoliberalizing space." *Antipode* 34(3): 380–404. doi: 10.1111/1467-8330.00247

Peck, Jamie, Nik Theodore, and Neil Brenner. 2013. "Neoliberal urbanism redux?" *International Journal of Urban and Regional Research* 37(3): 1091–1099. doi: 10.1111/1468-2427.12066

Piore, M and J Sabel. 1984. *The second industrial divide: possibilities for prosperity*. New York: Basic Books.

Popkin, Susan J, Bruce Katz, Mary K Cunningham, Karen D Brown, Jeremy Gustafson, and Margery Austin Turner. 2004. *A decade of HOPE VI: research findings and policy challenges*. Urban Institute May 18. www.urban.org/url.cfm?ID=411002&renderforprint=1&CFID=78071151&CFTOKEN=47422001 &jsessionid=f03070a4cb957b9301cd80707de14266c247

Porter, Michael E. 1997. "New strategies for inner-city economic development." *Economic Development Quarterly* 11(1): 11–27.

Propublica. 2014. "Bailout recipients." http://projects.propublica.org/bailout/list

RealtyTrac. 2008. "U.S. foreclosure activity increases 75 percent in 2007." *RealtyTrac* January 29. www.realtytrac.com/content/press-releases/us-foreclosure-activity-increases-75-percent-in-2007-3604

Sagalyn, Lynne B. 2007. "Public/private development: lessons from history, research, and practice." *Journal of the American Planning Association* 73(1): 7–22. doi: 10.1080/01944360708976133

Sager, Tore. 2011. "Neo-liberal urban planning policies: a literature survey 1990–2010." *Progress in Planning* 76(4): 147–199. doi: 10.1016/j.progress.2011.09.001

Sassen, Saskia. 2012. *Cities in a world economy*. Thousand Oaks, CA: Sage.

Saxenian, AnnaLee. 1996. *Regional advantage: culture and competition in Silicon Valley and Route 128*. Cambridge, MA: Harvard University Press.

Scott, Allen J. 2000. *The cultural economy of cities: essays on the geography of image-producing industries*. Thousand Oaks, CA: Sage.

Smith, Neil. 2002. "New globalism, new urbanism: gentrification as global urban strategy." *Antipode* 34(3): 427–450. doi: 10.1111/1467-8330.00249

Smith, Robin E. 2011. "How to evaluate choice and promise neighborhoods." *Urban Institute* March 18. www.urban.org/publications/412317.html

Squires, Gregory. 1989. *Unequal partnerships: the political economy of urban redevelopment in postwar America*. New Brunswick, NJ and London: Rutgers University Press.

Sugrue, Thomas J. 2014. *The origins of the urban crisis: race and inequality in postwar Detroit*. Princeton, NJ: Princeton University Press.

Teaford, Jon C. 2006. *The metropolitan revolution: the rise of post-urban America.* New York: Columbia University Press.

US Department of Housing and Urban Development. n.d. "HUD Implementation of the Recovery Act." http://portal.hud.gov/hudportal/HUD?src=/recovery/about

White House. 2012. Neighborhood Revitalization Initiative. www.whitehouse.gov/administration/eop/oua/initiatives/neighborhood-revitalization

Wilson, William Julius. 1997. *When work disappears: the world of the new urban poor.* 1st Ed. New York: Vintage Books.

Wyly, Elvin K, Mona Atia, and Daniel J Hammel. 2004. "Has mortgage capital found an inner-city spatial fix?" *Housing Policy Debate* 15(3): 623–685.

Zukin, Sharon. 1989. *Loft living: culture and capital in urban change.* New Brunswick, NJ: Rutgers University Press.

5

Urban politics and development

As chapters 3 and 4 have explained, the interplay of global trends and federal policies affects urban development patterns and revitalization outcomes. Nevertheless, urban revitalization is not simply a product of external forces. People at the local level also influence how their cities and regions fare in the global economy and work to build places that reflect regional values and priorities. In other words, urban stakeholders both *respond to* the regional and global trends of which they are part and they *shape* how the global trends play out locally.

In a neoliberal, globalized, economically integrated world, local actors continue to influence development outcomes within the ebb and flow of market forces and the regional urban structure. Local actors and coalitions possess varied priorities and power. This too shapes revitalization proposals and outcomes. Most urban stakeholders are growth-oriented, as we will discuss in this chapter. Nevertheless, urban governing regimes also incorporate different priorities such as racial equality or environmental protection, and take action to effect various types of change. Moreover, urban regimes are evolving and challenged by a variety of actors who at times value priorities other than urban growth.

In short, the interplay of urban politics, local context, and global forces affects how and where urban revitalization programs get implemented. In this chapter, we focus on the politics of urban development. We first discuss the key actors involved in urban revitalization in the contemporary global, neoliberal era. Next we describe their involvement in urban regimes and the urban growth machine, concepts that help us to understand how local action occurs and how priorities in urban revitalization are formed. Following this, we turn toward a discussion of the ways in which the metropolitan urban structure in the US impacts local action. Finally, we briefly discuss the development process and the wide-ranging ways the private sector and local government influence development.

URBAN POLITICS: THE ACTORS

Urban politics refers to the actions of and negotiations among all the people collectively working to develop and redevelop urban areas. In chapter 1 we discussed the major sectors involved in urban development and revitalization. These include the *state*, the *market*, and the *third sector*, comprised of NGOs and nonprofit organizations. The *people* who live, work and play in an area are also important participants. Planners and redevelopment professionals can work in any sector.

They can work within public agencies, as consultants in the private sector, or for nonprofit organizations. Residents can also instigate revitalization activity. In practice, actors in each of these "sectors" are working for specific agencies with different charges, businesses with particular needs, and organizations with varying constituencies. Individuals from each sector bring a particular way of seeing to a discussion that reflects their expertise and priorities. As we discuss in the following section, actors in each of the sectors work together and against each other in urban regimes or governing coalitions to set urban priorities, solve problems, and get projects off the ground.

The local government and the business community are two of the most influential players in urban development. The local government or the public sector refers collectively to all government actors and agencies, from elected officials to capital facilities staff members and the public works department's engineers. Within the city, city council members and elected public officials as well as department heads and other upper-level administrators have the most influence on urban and economic development priorities and initiatives.

Given the importance of economic development and growth, business leaders are equally or, at times, more influential. The work of cities—from establishing public priorities, and determining public investment to solving urban problems, attracting outside private capital, and mobilizing local capital—involves collaboration between the public and private sectors.

Additionally, in many situations, local residents and community-based organizations (CBOs) advocate for or challenge revitalization projects. Their involvement can impact the direction that redevelopment takes as well. As the federal government reduced its involvement with social welfare, this changed the role of the third sector or nonprofits. Nongovernmental organizations have taken over many programs that were previously provided by government or helped to fill the gap in government support, particularly in areas such as social service provision, job training, and affordable housing. Proponents of this structure argue the nonprofits are closer to the communities that they serve and, as a result, provide needed services more effectively and in culturally appropriate ways. A more critical perspective refers to this arrangement as a shadow state. Observers have shown that nonprofit organizations have increasingly become reliant on governmental grants and provide what were formerly public services but they are not subject to the same level of public oversight (Wolch 1990: 6).

Although many nonprofits are CBOs and reflect community priorities, this is not always the case. For example, Enterprise Community Partners is one of the largest nonprofit housing developers in the US. It is a national "501(c)(3) charitable organization that provides expertise for affordable housing and sustainable communities." While the organization develops local partnerships when working in an area, it also draws on national resources and relationships (Enterprise n.d,). In recent decades, nonprofits also have become more dependent on philanthropic dollars and entrepreneurial activities to fund their activities. The funding sources influence organizations' activities when they pursue available funding, rather than further mission-oriented objectives. This threatens how nonprofit organizations contribute to civil society and mediate between the public and private sectors (Eikenberry and Kluver 2004). Board members of nonprofit organizations are chosen based on connections and skills. This also links nonprofits to the business community as well as their constituents.

All development influences people who live, work, go to school, recreate, or frequent an area for any reason as well as those who hold property in an area. Ordinary people both impact and are impacted by how cities change. In many urban redevelopment situations, the participants refer to citizen or public participation when they discuss the involvement of ordinary people. In these cases, citizens refer to all people who are impacted as ordinary people and, in most cases, they are participating because they are impacted by a proposal rather than in their professional role. A critical dimension of citizenship is that people have both a right to influence what goes on, and responsibilities for local stewardship and engagement. It is important to understand, however, that "citizens" refers to all people. It is not referring to national citizenship or people holding US citizenship or authorization to work in the US. Given national and global migrations and real estate as an investment, some stakeholders might have property in an area where they do not live. Others live where they do not own property.

Members of the public have widely different interests and perspectives, so much so that it is more accurate to speak of multiple publics rather than one singular public. Residents might have different priorities from absentee property owners, as well as from each other, and from the professional and business interests in the area. Racial and ethnic identities as well as language can influence who people consider their communities. People who own property will view new proposals through a different lens than those who do not. Children and young adults need different types of recreational facilities in parks than seniors. Ultimately, all revitalization efforts impact regular people whether or not they participate in the process. How redevelopment will impact the diverse public constituents for redevelopment is a foremost concern. In chapter 13, we discuss field methods to help evaluate how an area is used and who might be relevant stakeholders. In chapter 14, we specifically discuss participation by residents and other ordinary people who live, work, or play in the area.

GOVERNING URBAN DEVELOPMENT: REGIMES AND GROWTH MACHINES

Scholars have developed different frameworks to explain how and why people work together to guide urban development. Urban regime and growth machine theories are two of the most useful approaches. Both explain urban governing arrangements in the United States within a political economy framework. A political economy framework, as we discussed in previous chapters, incorporates both the global economy and the role of government at different scales. These theories help explain how local governments and other local actors act on the opportunities and constraints they experience.

Urban regime theory examines the relationships between local governments and other stakeholders. Stone (1989: 6) defines an urban regime "as the informal arrangements by which public bodies and private interests function together in or to be able to make and carry out governing decisions." He goes on to emphasize that urban regimes do not run government, but that they are involved in "managing conflict and making adaptive responses to social change." An urban regime is therefore a set of governing arrangements that cross sector boundaries. It describes how the local government works with business leaders and other urban stakeholders to define a city's priorities, and influences what gets built and where. Regimes are stable through time, but they also change in response to shifting conditions and new priorities.

The mayor and the city council members are the visible officials elected to influence a city's agenda. To some extent, elected officials are directly responsive to the issues and concerns raised by residents. However, they are also responsible for the city's budget and they therefore expend much energy creating conditions to maximize city revenues. Because high property values increase the tax base and new employers bringing jobs can increase population and local spending, mayors and council members often privilege economic development agendas over other urban objectives.

Because economic development is critical for cities, local business interests have significant influence in urban regimes. Since the enactment of municipal corporations in colonial US towns, business leaders have taken an active role in city governance. They have been community leaders and worked to create robust business environments. For example, in chapter 3 we discussed the role of downtown business and property interests in shaping urban revitalization policy in the face of postwar suburbanization. Often, business priorities have the most influence (Stone 1989). Business interests can include local entities as well as global or national corporations with local headquarters or manufacturing sites.

Third sector organizations, religious leaders, social movement organizations, and neighborhood organizations also have influence and participate in urban regimes. Through their informal relationships and networks as well as official decision-making processes, they engage in political lobbying and set priorities and mobilize resources for development and redevelopment (Mossberger 2009, Stone 1989). Particular leaders can be important advocates for a given project or agenda, but the coalitions are more stable than the individuals in office or leadership roles at a given time.

The most common and stable type of urban regime is a development or pro-growth regime (Hanson 2003, Mossberger 2009, Stone 1989, 1993). As the name implies, development regimes are focused primarily on promoting growth. Coalitions of public officials and business leaders work to attract new industry and create a pro-business environment. This may include tax or other development incentives, or assistance in acquiring and preparing land for real estate development. Growth coalitions have often worked to make large tracts of land available for urban renewal projects. Public officials may also establish special purpose districts or offices to realize their agenda.

Private sector individuals and organizations work closely with the public sector. For example, in Atlanta, an organization of large property owners in the downtown called Central Atlanta Progress focused its efforts on influencing public sector action to enhance downtown property values. In the 1980s its priorities for downtown included the development of new housing for wealthy residents and revitalizing the Underground, a once thriving entertainment district. To this end, the group participated in a central area study that provided the basis for new plans for the downtown, which included public subsidies for affluent housing and redevelopment. They also helped to establish a public–private partnership to facilitate the Underground Atlanta's revitalization, and joined with a coalition that included the mayor's office, the city council, and multiple other partners in a public–private partnership that oversaw downtown redevelopment (Stone 1989). However, despite the efforts of Central Atlanta Progress and its partners, including millions in public subsidies, the Underground was a risky project that never realized its potential. In the 2010s, the question regarding what to do about Underground Atlanta was yet again on the growth agenda (Bluestein 2012).

Although cities have countless examples of failed urban revitalization initiatives, the growth imperative appears so strong that Logan and Molotch (2007) argue that the city is a growth machine. They contend that virtually all place-based stakeholders are growth-oriented. Governmental officials, business leaders, institutions such as universities and museums, small business owners, and residential property owners all have an interest in seeing a city or region grow. Because all stand to benefit, when an area grows and population and property values increase, the stakeholders have a shared interest in urban growth even when they do not share the same interests.

Problems arise with growth agendas because growth coalitions privilege actions that raise property values and have the potential—often unrealized—to change local economic dynamics but do not take into account other social dimensions. Large redevelopment projects may displace residents, require large subsidies that do not generate new, quality jobs, and can threaten environmental quality and historic neighborhoods. Market-oriented approaches also provide the most benefits in areas with the greatest growth potential. As a result, they often do not benefit neighborhoods with the most need for reinvestment because these areas do not appear to have sufficient market value. Growth-oriented strategies can perpetuate or even increase social inequality.

Pro-growth regimes are not the only type of urban regime, however. In addition to development regimes, Stone (1993) has characterized other regimes as maintenance or caretaker regimes when the primary role is the maintenance of basic services and low taxes. Middle-class progressive regimes emphasize a nongrowth-oriented political agenda. Lower-class opportunity expansion regimes focus on access to employment and social services (Mossberger 2009).

Santa Cruz is an example of a city that has maintained a middle-class progressive regime for more than four decades. In this relatively small California city, the pro-growth coalitions have lost on virtually every growth-oriented proposal from highway expansions to a convention center since 1968. Since 1981, the city council was dominated by "a confederation of socialist-feminists, social-welfare liberals, neighborhood activists and environmentalists" (Gendron and Domhoff 2008: 3). The business community could not gain enough influence to implement a development agenda, and often lost in negotiations with environmentalists and neighborhood activists although they still have had success in furthering some development projects (Gendron and Domhoff 2008). At times, the social-welfare liberals in Santa Cruz would support new development that would create jobs and increase revenues for welfare programs.

Although Santa Cruz is an exceptional case, in many cities alternative development coalitions that advocate for environmental protection or neighborhood quality-of-life issues have successfully steered growth-oriented agendas to address other issues. Such challenges have altered the dominant political agenda to include a focus on amenities, culture, and quality of life and brought new actors into the governing process with more progressive development agendas (Clark 2004, Grodach and Silver 2013, McGovern 2009).

Additionally, stakeholders have at times organized to increase opportunities for lower-income residents. However, no cities have developed and retained a lower-class opportunity expansion regime. In other words, no city has primarily developed an urban agenda that focuses on increasing opportunity and equality for residents with lower incomes. Nevertheless, cities can take action to improve conditions for low-income neighborhoods. In Indianapolis during the 1990s, under the leadership of Mayor Goldsmith, the city realized that its future economic prosperity required neighborhood revitalization. Under the previous administration, the city had successfully revitalized the downtown core but neglected some of the most disinvested neighborhoods. Even though Mayor Goldsmith was focused on reducing the size of local government, he also invested in neighborhood revitalization in ways that could increase residents' capacity to improve conditions. He directed Community Development Block Grant (CDBG) funds to seven neighborhoods in the most need, giving local CBOs the decision-making power on program funding (for programs such as day care and neighborhood cleanup) and directed $60,000 to ten churches to maintain 20 neighborhood parks. The city also funded the Indianapolis Neighborhood Resources Center to provide comprehensive training and technical assistance for neighborhood groups (McGovern 2003).

Stakeholders in growth coalitions also can work towards different objectives at the same time as pursuing growth. Atlanta is a good example. In the decades after World War II, Atlanta transformed from a white-majority city to an African American-majority city. African American leaders saw benefits of growth, but they simultaneously retained a civil rights agenda. Atlanta developed a bi-racial pro-growth regime that negotiated positions that were both growth- and equality-oriented. In direct ways, African American leaders shaped how development played out. In the case of the Underground Atlanta redevelopment, the council insisted on significant minority-owned business participation including 25% of construction contracts. Four council members had also insisted that a majority of contracts go to Atlanta businesses, but upon opposition from the project backers a negotiated settlement only encouraged local residents be hired. Only the grassroots organization ACORN later protested when out-of-town construction firms were used (Stone 1989).

This example shows that Atlanta growth coalition actions incorporate multiple objectives while primarily privileging the interests of the business community. The regime came to understand that ensuring disadvantaged businesses had an opportunity to participate in the city's growth was good business and later took a favorable position on such policies (Stone 1989). In this way, the political agenda also changed how Atlanta did business.

It is important to recognize therefore that growth coalitions have incorporated social change agendas. In a more direct example, Atlanta's coalition of white and African American business and public leaders decided that desegregating schools peacefully was in the city's interests. They successfully set an agenda that opposed racial discrimination in schools which resulted in a relatively conflict-free process that contrasts starkly with other Southern cities such as New Orleans. This occurred because African American coalition members insisted that racial equality be part of the city's agenda. Through their actions, the phrase "the city too busy to hate" was coined in the 1950s to promote a progressive vision of a new Atlanta based on racial equality (Stone 1989, 2001). This slogan attracted African American middle-class residents from other parts of the US, leading to population growth and central city revitalization.

Toronto has also developed its identity around multiculturalism and prioritized inclusive racial and ethnic politics (Boudreau, Keil, and Young 2009). However, when facing choices that put growth and equity considerations at odds, Toronto regularly chose growth over social equality. Although Toronto embraced its diversity that came from immigration from numerous places over the decades, disparities in opportunity and income among different groups remain high (Boudreau, Keil, and Young 2009). Likewise, although Atlanta supported African American middle-class residents well, it did not eliminate poverty or change circumstances for poor neighborhoods that were disproportionately African American.

Race and racism—structurally embedded in both urban policy and individual actions—continues to pervade all types of urban development concerns in large cities (Boudreau, Keil, and Young 2009, Hanson 2003, Stone 1989, Thomas 2013). The negative impacts of urban renewal and redlining have been widely acknowledged and documented. Racism has also contributed to urban decline and made it more difficult to create effective agendas for urban revitalization (Sugrue 2014, Thomas 2013). When Thomas took an in-depth look at Detroit to determine how the city lost population and fell into such devastating decline despite the efforts of countless smart, hardworking people, she found a host of factors. Metropolitan urban structure, which we discuss below, hurt the central city when auto manufacturing moved to the suburbs and later left the city in response to national and global trends. Additionally, there was not a robust urban policy in place to help guide development or ameliorate the impacts of suburbanization. Equally important, however, was the individual and institutional racial bias that operated in all decisions and plans. This led to housing segregation, blocked various initiatives, and limited regional action (Thomas 2013). As this example shows, even though growth-oriented proposals gain more traction, social justice is also necessary for long-term urban sustainability.

However, the autonomy of US cities and regions makes this hard to achieve. In the US, governmental power is decentralized and federal urban policy is weaker than in many countries. State governments also have significant autonomy from the federal government. While decentralization reflects the American belief in local control, it also creates particular constraints for local actors. Because relatively few federal funds are directed to cities and there is no national planning for different regions, US cities are highly dependent on local revenues such as sales and property taxes. Because income sources come from local economic and real estate development, municipal governments and business leaders promote growth to increase revenues. Since spending patterns reflect residents' incomes as well as how many people live, work, play, and shop in an area, they privilege middle- and upper-income residents' concerns because they need those residents to increase or maintain high revenues.

URBAN POLITICS AND GOVERNANCE IN A GLOBAL, NEOLIBERAL ERA

Although pro-growth market orientation in urban politics began long before the contemporary global neoliberal era, neoliberal changes affect local action. Early boosters tried to situate their cities to become centers of commerce as well as to attract more residents. In the nineteenth century, cities such as Chicago, Detroit, and New York established themselves as centers of manufacturing with a global reach while many other places vied to become railroad hubs or homes to universities or even prisons (Logan and Molotch 2007). In cities such as Dallas, business leaders held far more power than government officials because local political culture emphasized small government and limited regulation under an entrepreneurial regime that governed for decades (Hanson 2003).

Nevertheless, the second half of the twentieth century brought significant changes to how industries function locally. This generation of business executives, now corporate CEOs rather than local business leaders, are less interested in the affairs of a given city. Their offices are as likely to be in a corporate office park as a downtown office building. Their attention is focused on global markets rather than local politics. As corporations have grown and globalized their manufacturing processes and management, they have become less place-dependent. This reduces local attachment because without local ties corporations can more easily move away from the region. Paradoxically, this also gives corporations more relative power in local negotiations. Influential corporate decision-makers no longer live in all the cities where they operate. Formerly independent companies are now sometimes part of national and international corporate conglomerations. In the process, business leaders have become less invested in any one location (Hanson 2003).

The globalization and corporatization of many industries has created different constraints on a city's actions, but the local and national context can alter how this plays out. According to Savitch and Kantor (2002), US cities suffer from low federal involvement and weak urban policy because this forces them to be more competitive and have less bargaining power when engaging with corporations and non-local actors. Weak-market US cities are the most fragile, and residents end up with worse outcomes compared with weak-market cities in other countries with higher levels of national intervention and stronger urban policies.

It is also important to consider how new actors have gained influence in this environment. For instance, according to Hackworth (2007: 19), "bond-rating firms, such as Moody's Investors Service and Standard and Poor's (S&P), are perhaps the single most influential institutional force in determining the quantity, quality and geography of local investment." This matters for cities because most do not have adequate surplus to invest in long-term development. The revenues in a city's general fund primarily support day-to-day operations. They pay for city staff, the police and fire departments, and other daily operations. As a result, they borrow money to build infrastructure or invest in large-scale economic development.

The two common forms of municipal long-term debt are general obligation bonds and revenue bonds. In the first case, the debt is repaid through the general revenue and the bonds are usually subject to a referendum. Revenue bonds are tied to specific revenue streams such as an airport-user fee and are not usually subject to referendum. Bond-rating agencies become important because they rate the municipality's credit-worthiness. If a municipality has a high rating, the investment risk is lower and the bonds are sold more easily to investors and investment banks. Because a city's credit-worthiness is dependent on its overall revenue flow and expenses, retaining a high credit rating impacts other city functions. Providing social services, recreation, and other activities becomes more difficult because they add expenditures without generating direct revenue (Hackworth 2007).

As we discussed in chapter 4, one effect of market-oriented neoliberal ideology has been to reduce governmental programs and social welfare functions. This roll-back period has had profound effects on cities (Peck and Tickell 2002). When the federal government reduced its commitment to end poverty and improve urban conditions, federal sources of revenue declined, making cities even more dependent on the private market. When the federal government reduced its social service provisions—previously implemented through initiatives such as the War on Poverty—those obligations were shifted to local authorities or state governments. However, because cities are already searching for new sources of revenue and their bond rating has become more important in this regard, social services have become more limited at the local level as well. Fiscal austerity therefore has come at the expense of improving opportunity and life quality for many urban residents.

Frequently, cities try to offset the costs of providing an amenity-rich urban environment by approaching amenities as an economic development strategy to attract wealthy residents and business activity. This reinforces the incentive for cities to work to attract mobile, high-income residents. Over the past decade this strategy has become commonplace as public officials have been influenced by Richard Florida's (2002) creative class theory. Florida argues that members of the creative class—artists, designers, engineers, and an array of residents with high levels of human capital—increasingly make decisions about where to live based on quality of life considerations rather than employment opportunities. Accordingly, because the creative class seeks places with a high level of recreational and cultural amenities, cities have invested in items such as bike paths and new museums in order to attract these mobile workers. Indeed, some argue that the role of amenities and consumption has taken a more central focus in the agendas of urban regimes causing them to shift their role from growth machine to entertainment machine (Clark 2004). However, although cities often claim that all residents benefit and the improvements are politically neutral, in practice these improvements often reflect the amenity priorities of wealthier workers over affordable housing, opportunities for youth, or other neighborhood concerns.

New York has been a visible example of a city recreating itself to suit the interests of wealthier residents. New York's Mayor Michael Bloomberg and his administration helped make the "CEO mayor" an accepted model for urban leadership. The administration's rhetoric focused on efficiency and political neutrality, and suggested the city would be run like a business. This vision was contrasted with the partisan or special interest political image. The "neutral" position, however, obscured the effect of the policy and public investment decisions that intended to develop a luxury city for upper-class global elites who lived, worked, and shopped in New York (Brash 2011).

For instance, the Bloomberg administration supported a private plan put forward by a group of business leaders for Hudson Yards even as a coalition that included block associations, community groups, and affordable housing advocates offered alternatives that better retained the area's scale and character. The neighborhood's vision, defined in the Special Clinton District Plan, included a mix of residential and industrial uses, affordable housing, waterfront access, traffic mitigation, and high environmental standards for neighborhood residents. The city developed a plan that paralleled the objectives originally put forward by a group of business leaders that envisioned dense commercial and residential development hotels with public space designed in conjunction with the commercial development. The neighborhood coalition also rejected rebranding the area as Hudson Yards or Midtown West and instead preferred Hell's Kitchen South to retain the connection with the neighborhood's diverse uses and history. Although Bloomberg's administration purportedly was not focused on elite interests and instead emphasized good planning and economic benefit, and it initially responded to neighborhood concerns, the administration ultimately supported high-density, upscale residential and commercial uses for Hudson Yards and self-financing mechanisms that reflected the vision that the neighborhood residents had opposed. Bloomberg's administration therefore was not apolitical but instead focused on the interests of the global business class.

COMMUNITY DEVELOPMENT AND ORGANIZING

In neighborhoods that have suffered from disinvestment, local organizations and residents spur community development initiatives to make change in the local area. The purpose of community development is to create economic opportunities and civic and leadership capacity while also improving neighborhood conditions. Residents can organize collective action to respond to numerous conditions. Residents have responded to the gambit of local challenges from vacant property to the need for affordable housing, job training, small business support, childcare or local services, or grocery stores.

As much research has shown, however, "the problems of inner-city neighborhoods derive from structured social relationships that deplete the resources of the community and exclude residents from control of both capital and decision-making" (Saegert 2006: 277). An important part of changing neighborhoods' and residents' circumstances is organizing residents to achieve diverse goals. To address disinvestment and related problems for urban residents, people have engaged in community development, community building, and community organizing. Community development often operates through CBOs and emphasizes developing human capital and social capital. Community building emphasizes building relationships, coalitions, and capacity to effect change (Saegert 2006). Community organizing is related to community development and community building, but describes a power-oriented form of political organizing where, through collective action, residents make demands for change and develop the capacities to make that change happen. We discuss community organizing in chapter 14.

Local initiatives often combine efforts to build capacity with material, neighborhood change. Community development corporations or CDCs are community- or neighborhood-based nonprofit organizations that are founded to stimulate local revitalization. They are usually organized by local stakeholders including residents, community and religious leaders, and local businesses. They work to stimulate investment and build relationships with corporate and business leaders who can help bring quality jobs to an area, or who can help fund other initiatives. Many CDCs focus on affordable housing or business support, but across the US, CDCs are diverse organizations that respond to the conditions in the areas they serve (University of Alabama Center for Economic Development 2013).

One prominent example, Dudley Street Neighborhood Initiative (DSNI), has been successful at greatly improving conditions in Boston's Dudley Street Neighborhood for 30 years. DSNI's mission is "to empower Dudley residents to organize, plan for, create and control a vibrant, diverse and high quality neighborhood in collaboration with community partners" in a neighborhood that was once threatened by arson and dumping and where residents lived with more abandoned property than vibrant parks. DSNI has prioritized revitalization without displacement and worked to build affordable housing and to bring fresh food and local economic development as well as neighborhood improvement (www.dsni.org). Cities across the US have organized neighborhood-based initiatives and CDCs that focus on renewal of specific neighborhoods for the residents who live there.

URBAN POLITICS AND GOVERNANCE IN METROPOLITAN AMERICA

Communities, neighborhoods, and cities themselves are also situated in a metropolitan context. Metropolitan areas have numerous local jurisdictions along with unincorporated suburbs, county governments, regional planning bodies, and a plethora of special districts to provide urban services. As early as the 1920s, metropolitan areas grew and the suburbs increasingly chose to remain independent or to incorporate as separate cities. These circumstances created politically fragmented regions that define today's urban political landscape (Teaford 1979). In 2015, Los Angeles County included 88 distinct cities. Beverly Hills and Santa Monica, for example, are surrounded on all sides by the City of Los Angeles.

Independent municipalities have the authority to tax, provide services, and allow or disallow different types of development. Residents make decisions where to live and businesses decide where to locate because of the specific place differences and advantages over another within the region. The relative differences among jurisdictions within a region therefore affect where development occurs. When better-quality housing became available and affordable in the suburbs, people with resources moved out of the central city. Historically, white, native-born residents developed restrictive covenants to prevent people of different races and ethnicities from moving to the neighborhood. They also had the ability to prevent industry from moving in or to require houses be built on large lots. As a result, cities have different character from one another and they have different levels of affordability and amenities.

People cross municipal boundaries when commuting to work and businesses make location decisions to take advantage of region-wide opportunities and differences among them. The fragmented political jurisdictions both have contributed to and reflect uneven patterns of development. Suburbanites could benefit and work in the central city while retaining political independence—and that included overt attempts to enforce racial hierarchies—in residential neighborhoods. Increasingly, jobs also moved to suburban areas, further separating the lives of people from the urban core. This highly localized autonomy, combined with discriminatory individual actions and public policies, created patterns where some areas have wealth and some have greater poverty. It also created disincentives for wealthier areas to think regionally.

The desire to live with people who had similar incomes was an important driver behind local control in the form of a fragmented metropolis. Property and sales tax revenues are higher per capita in high-income areas. When spent locally they create better parks, schools, and street paving. This also creates an incentive to keep property taxes high and keep out low-income residents. To counter this tendency, some states such as California, New Jersey, and Washington have created fair share housing legislation to ensure all jurisdictions provide some affordable housing. Because residents can "vote with their feet" or move to a city within the region that provides the best bundle of services and amenities, the metropolitan structure makes it difficult for municipal governments to redistribute resources to residents who have greater need instead of providing services for those who pay more taxes (Peterson 1981). Because cities compete with one another for both residents and business in order to increase their revenues, they orient themselves to development and revitalization policies that will attract wealthy residents, businesses, or tourists rather than improving the lives of all residents. As we mentioned earlier, US cities have significant autonomy so they can decide how to work with nearby jurisdictions or if they even want to do so. In some cases, cities that are relatively wealthier see little reason to work with cities in need of more revitalization.

The federal government can also hold large swaths of land such as military bases that influence local politics. In some areas, tribal governments have interest in local redevelopment efforts when they hold land or their members live in an area. They also might have interest in how new development will impact water or other resources, and they engage in economic development activities that can impact communities off reservation. Because they have national sovereignty rather than functioning as a local jurisdiction, they participate differently and have different types of autonomy to take action than local agencies.

THE LOCAL GOVERNMENT AND DEVELOPMENT DYNAMICS

The local government plays a role in all development projects, whether they are instigated by professional developers, incentivized by public agencies, or spurred by community residents. In the US, we often consider our governments democratically elected and the policymakers accountable to the public constituencies, but the realities are more complex. Governmental agencies have differing priorities, many agencies are guided by state and federal laws, and some constituents have greater influence than others. Although the elected officials help set priorities, the city's budget, its municipal codes, strong private sector and community group interests, and various other factors influence what government can do.

Limited municipal budgets are another important constraining factor. Cities appear to have large budgets, but very little of the recurring revenues are discretionary. In 2014, the City of New Orleans had an operating budget of approximately $500 million. This needs to fund everything from police officers to street light repairs. In New Orleans, fire and police service alone costs over $200 million and sanitation costs close to $40 million. When the countless other services are included—permitting, planning, public works, municipal and traffic court, parks and recreation—the funds available for new initiatives are limited. Any general revenue bond debt would also come from the general fund. Another serious consideration for many cities—particularly those with stagnant and slow growth economies—are defined benefit retirement plans. When employees retire, they receive a defined amount for the rest of their lives. In 2010, New Orleans spent $78 million on pensions. As people live longer, pension obligations have grown, in many cases, even as the workforce paying into the system has decreased (BGR 2012).

Municipal revenues fund the myriad agencies that influence the development process. In this sense, municipal governments are partners in all development that occurs. It is important to remember that real estate development, local commercial activity, streets and public spaces, and even individual residences are highly regulated. Local, state, and federal governments influence virtually all aspects of urban life—that which occurs formally, within the bounds of authorized markets and processes, and informally; or that which occurs without authorization. Regulation is so pervasive that conservative and liberal critics alike find some (but not the same) regulation unjust and onerous. Although in the neoliberal era some functions were deregulated, or the regulations were reduced, it is more accurate to say that we have different rather than less regulation. The state shifted its regulations from mediating market failures to protecting and facilitating economic growth.

City residents influence the development process through the city council as the elected officials enact ordinances that establish guidelines for development, public use, and a host of other activities that occur. Although some appear technical—for example earthquake-resilient building standards in California—others, such as tree provision or design guidelines, can change the character of a city or neighborhood. Once the regulations are adopted and codified, they are reviewed and enforced by numerous agencies. The planning department reviews development proposals and works with different stakeholders to develop and modify planning documents that guide development. Public works departments, capital facilities planners, and parks departments have plans, investment, and maintenance priorities. Specific projects might be subject to design review or review by a historic preservation commission.

Although the public sector acts on behalf of "the public interest," in reality there are many, often conflicting *public interests* at play. Some priorities are shaped by federal and state policies and others are negotiated locally. Unexpected differences can occur among agencies that have different priorities and perspectives. Specific affordable housing and environmental regulations and plans are framed at the state level. The state government might have programs that influence available revitalization funding. The state policies also must be consistent with federal policies and with directives (or laws) and financing tools. These too have local impacts.

In addition to local government, universities, hospitals, and other nonprofit institutions can engage in redevelopment. The great majority of revitalization, however, is through private development. Real estate development is an important aspect of what binds city coalitions. Real estate value in the United States was estimated at $30 trillion in 2007: two-thirds was residential development and one-third was commercial development (Miles et al. 2007).

Developers have different motivations and ways of working from the public sector. Developers or a development team usually refers to private market professionals who make profit (or at least get paid) through designing and constructing as well as redeveloping the built environment. From the developer's perspective, a critical dimension is whether the project will be feasible and therefore profitable. This of course has implications for urban revitalization in the public interest. The first three steps of an "eight-stage model of real estate development" involve project feasibility (Miles et al. 2007). The first step is the evaluation of an initial idea for feasibility. If it appears feasible, the idea is refined and re-evaluated for feasibility. The third step is commissioning or conducting market and feasibility studies. If, at that point, the project still appears viable, it will move towards contract negotiations, then a formal commitment where agreements are formalized. After all of this, a project goes on to construction. The final two steps are completing the project and then managing the project or portfolio. The last step might involve the new owner if the project is sold (Miles et al. 2007).

Given limited budgets, many revitalization projects, in both weak- and strong-market regions, require some form of subsidy to make the project viable. These subsidies often come from federal programs including the CDBG Program, HOPE VI, and the Empowerment Zones as well as the Low Income Housing Tax Credits (LIHTC), Historic Preservation Tax Credits, and other available programs we discuss in section III. Although subsidies come from the federal government, they are often administered locally and cities can further particular objectives through the way in which they direct subsidies. For example, cities can use CDBG funds as layered subsidies for LIHTC projects to ensure that enough housing is built for people with very low incomes or special needs.

All development has end users and a neighborhood context. Satisfying the end users is a primary objective because the project's profitability depends on future buyers or tenants. Most conflicts arise with the neighborhood because even if a project is feasible, local residents might have different perspectives about its benefits or adverse effects, which we will discuss more in chapter 14.

SUMMARY

Urban politics describes how the public and private sector work together, more often than not, to promote growth in the city or region. Nonprofits and CBOs as well as citizens also shape the process, but the most influential decision-makers are most often coalitions of influential business leaders working with the public sector. All of these local actors both influence and are influenced by global forces and national trends.

The public sector shapes private development in many ways although it directly develops only a small part of all urban redevelopment projects. Public officials work in coalitions that set priorities and acquire resources and support to get large projects off the ground. All development—large and small—is also subject to municipal review and regulations. By enacting regulations, city residents through their elected officials set guidelines about the qualities and standards for new development.

Local processes shape how development occurs, but it is important to keep in mind the balance and negotiation between nonlocal forces and regional structure as well as values and priorities in local areas. Most urban actors are growth-oriented, but growth is not enough. For cities to become more sustainable, equity—in terms of both racial justice and economic opportunity—must also be a factor.

STUDY QUESTIONS

1 Why are cities focused on growth? What alternatives exist for weak-market cities?
2 Why are coalitions of public and private actors the common governing frameworks in the US?
3 What is the role of the nonprofit or third sector in growth coalitions? Why do they not play a stronger role?
4 What are some of the ways in which local government influences the development process?

REFERENCES

BGR. 2012. *The rising cost of yesterday: metro area pension costs and the factors that drive them.* BGR Pension Series. New Orleans: Bureau of Governmental Research.

Bluestein, Greg. 2012. "City to tackle the Underground dilemma again." *Atlanta Journal-Constitution* December 31. www.ajc.com/news/business/city-to-tackle-the-underground-dilemma-again/nTjMd/

Boudreau, Julie-Anne, Roger Keil, and Douglas Young. 2009. *Changing Toronto: governing urban neoliberalism.* Toronto: University of Toronto Press.

Brash, Julian. 2011. *Bloomberg's New York: class and governance in the luxury city.* Vol. 6. Georgia: University of Georgia Press.

Clark, Terry Nichols. 2004. *The city as an entertainment machine*. Vol. 9. Amsterdam: Elsevier/JAI.

Eikenberry, Angela M and Jodie Drapal Kluver. 2004. "The marketization of the nonprofit sector: civil society at risk?" *Public Administration Review* 64(2): 132–140.

Enterprise. n.d. "About Us." www.enterprisecommunity.com/about/mission-and-strategic-plan

Florida, Richard L. 2002. *The rise of the creative class: and how it's transforming work, leisure and everyday life*. New York: Basic Books.

Gendron, Richard and G William Domhoff. 2008. *The leftmost city: power and progressive politics in Santa Cruz*. Boulder, CO: Westview Press.

Grodach, Carl and Daniel Silver. Eds. 2013. *The politics of urban cultural policy: global perspectives*. Routledge Studies in Human Geography. Abingdon, UK and New York: Routledge.

Hackworth, Jason. 2007. *The neoliberal city: governance, ideology, and development in American urbanism*. New York: Cornell University Press.

Hanson, Royce. 2003. *Civic culture and urban change: governing Dallas*. Detroit, MI: Wayne State University Press.

Logan, John R and Harvey L Molotch. 2007. *Urban fortunes: the political economy of place*. Berkeley, CA: University of California Press.

McGovern, Stephen J. 2003. "Ideology, consciousness, and inner-city redevelopment: the case of Stephen Goldsmith's Indianapolis." *Journal of Urban Affairs* 25(1): 1–26.

McGovern, Stephen J. 2009. "Mobilization on the waterfront: the ideological/cultural roots of potential regime change in Philadelphia." *Urban Affairs Review* 44(5): 663–694. doi: 10.1177/1078087408323943

Miles, Mike E, Gayle L Berens, Mark J Eppli, and Marc A Weiss. 2007. *Real estate development principles and process*. Washington, DC: Urban Land Institute.

Mossberger, Karen. 2009. "Urban regime analysis." In *Theories of urban politics* 2d Ed., edited by J Davies and D Imbroscio, 40–55. London: Sage.

Peck, Jamie and Adam Tickell. 2002. "Neoliberalizing space." *Antipode* 34(3): 380–404. doi: 10.1111/1467-8330.00247.

Peterson, Paul E. 1981. *City limits*. Chicago, IL: University of Chicago Press.

Saegert, Susan. 2006. "Building civic capacity in urban neighborhoods: an empirically grounded anatomy." *Journal of Urban Affairs* 28(3): 275–294.

Savitch, Harold V and Paul Kantor. 2002. *Cities in the international marketplace: the political economy of urban development in North America and Western Europe*. Princeton, NJ: Princeton University Press.

Stone, Clarence N. 1989. *Regime politics: governing Atlanta, 1946–1988*. Lawrence, KS: University Press of Kansas.

Stone, Clarence N. 1993. "Urban regimes and the capacity to govern: a political economy approach." *Journal of Urban Affairs* 15(1): 1–28.

Stone, Clarence N. 2001. "Civic capacity and urban education." *Urban Affairs Review* 36(5): 595–619.

Sugrue, Thomas J. 2014. *The origins of the urban crisis: race and inequality in postwar Detroit*. Princeton, NJ: Princeton University Press.

Teaford, Jon C. 1979. *City and suburb: the political fragmentation of metropolitan America, 1850–1970*. Baltimore, MD: Johns Hopkins University Press.

Thomas, June Manning. 2013. *Redevelopment and race: planning a finer city in postwar Detroit*. Detroit, MI: Wayne State University Press.

University of Alabama Center for Economic Development. 2013. *Community Development Corporations Information Guide*. Tuscaloosa, AL: University of Alabama.

Wolch, Jennifer R. 1990. *The shadow state: government and voluntary sector in transition*. New York: Foundation Center.

SECTION III

Strategies, Policies, and Projects
The Spaces of Revitalization

In section III we turn our attention toward the application and outcomes of a diverse set of urban revitalization policies and strategies applied by local governments, public–private partnerships and community organizations in different contexts. The chapters contextualize the diverse trends that result in revitalization efforts and include case studies to illustrate how different projects unfold, highlighting the application of urban theory, policy, and politics.

We begin with a focus on revitalization planning in four specific contexts: downtowns, neighborhoods and housing, suburbs, and shrinking cities. In chapter 6 we begin with a discussion of the motivations and range of actions taken to reinvent downtowns. In chapter 7 we concentrate on neighborhood revitalization actions with a specific focus on the important role of housing. Following this, in chapter 8 we focus on efforts and techniques to revitalize inner suburban areas. This encompasses the challenge of "retrofitting" suburbia to new land use and design models, and the role of demographic changes reshaping traditional suburban areas. In chapter 9 we examine the emerging strategies to re-envision shrinking cities, cities that have sustained population loss for decades. In chapters 10 and 11, the final two chapters in this section, we concentrate on two important areas of revitalization applicable to each of these situations: efforts to create cleaner and greener urban environments, and those focused on rebuilding people-oriented spaces.

6

Reinventing downtown and the urban core

Twenty-first-century downtowns and surrounding neighborhoods have been in a difficult situation. What was once the central business district (CBD) is no longer the center of employment. Most major metropolitan regions now contain multiple business districts and dozens of municipalities that vie for residents, jobs, and commercial activity. Yet the downtown continues to be symbolically important within the region in the competition for global capital, and growth coalitions there work to retain their prominence as the urban center in the face of suburban growth. Downtowns in small cities also must compete with regional malls and suburban development, while seeking to preserve their identity as the city's symbolic center.

To survive and thrive in these circumstances, downtown growth coalitions have undertaken immense efforts to recreate the urban core around entertainment and consumption. Through new development and adaptive reuse, cities have invested in new flagship museums, sports stadiums, and convention centers. They have transformed disused warehouse and historic districts into new arts and entertainment zones while bolstering residential development and improving public transit. Derelict brownfield sites and abandoned industrial infrastructure have become public spaces and parks through both public and private efforts. Smaller cities have also incentivized downtown loft apartments or condos in underused warehouses and revitalized main streets by investing in improvements to support restaurants, specialty shops, and cultural institutions.

The redevelopment of central city commercial districts, waterfronts, and industrial spaces has successfully transformed many disused and dilapidated areas into destinations for middle- and upper-income residents, tourists, and conventions. They also provide an important source of jobs for many people.

These revitalization efforts have not come without costs, however. In chapter 1 we explained that the goal of urban revitalization is to rebuild cities in ways that diverse people can thrive and prosper. For this to occur, urban revitalization efforts must contribute to urban change that accomplishes the six objectives that we outlined: building human capital, promoting social–cultural equity, developing a desirable built environment for diverse groups, designing attractive public spaces, fostering economic competitiveness, and pursuing environmental sustainability. However, the goals of downtown revitalization are often driven and motivated by business and real estate interests seeking to protect and grow their investments. Their efforts can and do support some of these objectives, but they also have contributed to displacing residents as well as industrial and commercial establishments that conflict with the new vision of recreational consumption. While this has produced place-based improvements, it does not address long-standing urban problems and engender robust revitalization. In some cases, cities have attempted to balance the negative outcomes of development through community benefits agreements and linkage policies, but more work needs to be done to ensure that redevelopment both builds human capital and promotes social–cultural equity.

In this chapter we begin with a discussion of the changing character of downtown development. As we discuss, local governments and other central city interests have relied on a range of strategies related to urban design, zoning, land-use mix, and place image. They concentrate on inducing business and property development in the urban core by lowering development costs through a variety of tax abatement programs as well as the acquisition, assembly, and development of property. We then detail two strategies that epitomize the incentive- and place-based approach that has come to define downtown redevelopment: business improvement districts and tax increment financing. None of these strategies are exclusive to downtown revitalization and, while they have since been applied in varied contexts, they have been central to achieving the changes we have witnessed in US downtowns over the past decades. We close the chapter with a discussion of linkage policies and alternative development strategies that some cities employ to balance the negative outcomes of revitalization efforts and develop human capital through job creation for local residents.

THE CITY OF CONSUMPTION

With few exceptions, downtowns in major US cities no longer contain the majority of office space in the region, and therefore no longer function as central business districts. Atlanta, Dallas, Denver, Detroit, Houston, Los Angeles, Miami, and Washington, DC each contains 30% or less of its region's office space today. Chicago and New York stand out as the only exceptions, where the urban core barely contains a majority of office space in the region (Lang and LeFurgy 2003).

Instead, over the past decades, downtowns have been reinvented as centers of consumption (Clark 2004, Glaeser, Kolko, and Saiz 2001, Hannigan 1998, Judd and Fainstein 1999). Since the 1990s, large-scale projects and mixed-use districts have defined downtown redevelopment. They contain flagship museums and performing arts venues, themed restaurants and specialty retail, hotels, condominiums and loft-style housing, farmers' markets, convention centers, and sports arenas. Historic warehouse and factory buildings have been transformed to accommodate the new uses alongside new designer buildings. These places are marketed as fun, mixed-use, and pedestrian-friendly consumer destinations.

As we explain in chapter 2, three factors account for the rise of consumption-based development in the urban core. First, the growth of interurban competition and the dispersal of business functions shifted the focus of downtown development coalitions from attracting businesses to attracting suburbanites back to the central city. Second, downtown consumption-based development results from industrial restructuring where financial services and technology sectors locate in the urban core and bring a workforce with strong discretionary spending who often prefer urban living. Third, more young adults and empty nesters are attracted by the amenities and aesthetics of center city neighborhoods and opt to move there.

Vibrant downtowns are built around a mix of new and historic spaces in an attempt to create distinct urban experiences. The urban experience combines appealing industrial-era architecture with cutting-edge culture and the contemporary amenities deemed necessary to attract residents and visitors. Former industrial spaces have become premier shopping and tourist destinations. They evoke the city's past, and use appealing historic buildings and urban spaces to create distinctive identities (Frieden and Sagalyn 1991). Faneuil Hall in Boston, South Street Seaport in New York, Baltimore's Inner Harbor, San Francisco's Ghirardelli Square and Fisherman's Wharf, and Chicago's Navy Pier provide the early models for using historic and industrial architecture to create popular destinations (Figure 6.1). These continue to be major tourist draws that bring in significant revenues for these cities and have inspired others around the world seeking to replicate

Figure 6.1   Pier 39, Fisherman's Wharf, San Francisco, CA
Photograph by John O'Neil

their success. However, as these spaces are managed for visitors, critics charge that they are themed and sanitized centers of consumption. These insulated "tourist bubbles" conceal from visitors the urban poverty, crime, and homelessness that surround them, or displace these problems to other parts of the city entirely (Hannigan 1998, Judd 1999).

Downtown development has also turned toward the construction of large-scale cultural institutions. This strategy dates back to the 1950s when a coalition of civic and business leaders, under the leadership of John D Rockefeller III, sought to develop a new performing arts complex in New York City. Part of Robert Moses' plan for redeveloping the Upper West Side of Manhattan, Lincoln Center for the Performing Arts was the centerpiece of an urban renewal project created expressly to attract higher-end development to the area. Lincoln Center and subsequent 1960s-era cultural complexes such as the Kennedy Center in Washington, DC and the Los Angeles County Performing Arts Center were designed as fortress-like cultural compounds so that visitors could come and go without experiencing the surrounding area (Grodach 2011). Today, many cultural institutions have attempted to become larger tourist draws in part by turning toward spectacular, eye-catching architecture (Figure 6.2).

The urban experience extends to the diverse neighborhoods that make up the urban fabric. Ethnic neighborhoods and neighborhoods with rich histories have become an important part of urban living. The variety of food and culture contributes to giving the urban core its cosmopolitan character and sets it off from the surrounding suburbs (Zukin 2010). The growing appeal of urban core living also has encouraged shrinking cities to create initiatives to attract suburban residents back into the city. Cleveland for instance promotes itself for its "grit mixed with sophistication … where you can eat bucatini pasta served with beef jerky, dance to world music on the front lawn of a renowned art museum or do yoga in front of the Rock and Roll Hall of Fame" (Positively Cleveland 2014).

Often this activity is centered around a district modeled on New York's SoHo which offers art galleries, restaurants, and night clubs in historic industrial spaces. Although people experience

Figure 6.2   Aerial view of Experience Music Project, Seattle, WA
Photograph by SFM Archive, under www.gnu.org/copyleft/fdl.html

such neighborhoods as "organic" development, in most instances redevelopment has been encouraged through public policy initiatives and zoning strategies. Through both public and private efforts, places such as downtown Los Angeles, uptown Oakland, the Crossroads Arts District in Kansas City, and Wicker Park in Chicago have been recreated as authentic urban neighborhoods.

However, some fear that downtown redevelopment processes result in the suburbanization of the urban core itself (Hammett and Hammett 2008). Even as these neighborhoods are marketed for their urban authenticity, they are increasingly occupied by corporate and boutique retail, and patrolled by private security (Chapple, Jackson, and Martin 2010, Lloyd 2010) (Figure 6.3). In 1990, 55% of New York's SoHo neighborhood businesses were arts-related and only 10% were corporate retail and service chains. By 2005, 52% were corporate retail and services and only 9% were arts-related businesses (Zukin 2008). Even urban neighborhoods with identities rooted in the history of immigration, political activism, and ethnic ties focus on attracting outsiders, forcing them to respond to policy objectives and outsider concerns rather than responding to priorities of their residents (Dávila 2004).

These trends in downtown redevelopment bring two tensions to the forefront. Cities are attempting to create distinctive spaces, but in doing so they are replicating other cities. Although still appealing to many people, the experiences in these places become increasingly homogenized, even as the specifics—the markets, the architecture, the waterfronts, the neighborhoods—are rooted in a city's unique history and the new development incorporates innovative design. As a result, even as people want authentic urban spaces, the process by which they are created and changed can undermine the very qualities that people seek (Zukin 2010). Moreover, although individuals want to visit ethnic neighborhoods, they do not want to visit an ethnic neighborhood with other tourists like themselves.

The other tension revolves around the fact that much downtown redevelopment is the product of neoliberal policies meant to encourage gentrification (Hackworth and Smith 2001). Downtown

Figure 6.3 SoHo boutique, New York City
Photograph courtesy of Peter Blazak photography

redevelopment is a market-driven process stimulated partly by public incentives that have primarily benefitted central city property investors while eroding any possibility of achieving social and economic equity. Few projects have engaged with longtime community members, instead privileging private sector interests. In addition, in their efforts to attract development, public entities ease planning and zoning restrictions and offer tax and land giveaways that help large, corporate interests at the expense of small and local businesses (Fainstein 2005). Although some projects promise social concessions, the outcomes are market-driven and therefore geared toward tourists and middle- and upper-income residents.

Despite the tensions, entertainment and consumption define the new city center and provide a safe investment environment to attract people and capital downtown (Frieden and Sagalyn 1991, Leinberger 2009). In the process, consumption districts have brought new retail and entertainment venues, enabled the renovation of historic buildings, and improved the built environment while providing new sources of sales and property tax revenues. At issue, however, is the fact that they build a city for consumers and not diverse publics. In particular, these projects may not benefit lower-income urban residents, many of whom lived in the urban core prior to the period of gentrification. Although some projects bring important social benefits including affordable housing, employment opportunities, and retail, too often formerly struggling areas are transformed without making space for the diverse people who called central cities home. While this has produced place-based improvements, it often displaces problems rather than engendering robust urban revitalization.

REVITALIZATION TOOLS AND STRATEGIES IN DOWNTOWN AND THE URBAN CORE

To stimulate downtown revitalization, the most common approaches rely on tools intended to spur land development in specific places. As we discussed in chapter 5, downtown growth coalitions that include public and private sector participants decide what areas will be targeted for development. Business leaders advocate for and public entities provide incentives to reduce development costs. The incentives range from various forms of tax abatement to the public provision of new infrastructure and streetscape improvements. In conjunction with strategies that directly lower development costs, cities can revise and streamline zoning codes as well as reinvest tax dollars into the area through tax increment financing (TIF) districts or allow local businesses more controls through a business improvement district (BID).

In this section we discuss some of the most widely used incentive, land-use, and zoning strategies applied in revitalization efforts in downtowns and the surrounding urban core. We devote special attention to two established yet controversial tools: business improvement districts and tax increment financing. These strategies have been used and adapted in a variety of urban contexts, from urban cores in shrinking cities like Baltimore and Detroit to growing exurban municipalities in Texas. While these efforts can have a positive impact on the jobs and business development, revenue capacity, and image of central city areas, this can come at the expense of social and economic equity. In closing, we examine the ways that some cities have attempted to mitigate or balance the negative side-effects of their redevelopment efforts in the downtown and urban core.

PROPERTY-BASED INCENTIVES

Property-based incentives are some of the oldest and most ubiquitous tools that municipalities use to direct development into a particular neighborhood or part of the city. If you recall from chapter 3, redevelopment agencies used urban renewal funds for land assembly, to clear land, and to develop infrastructure in specific areas to induce development interest there. In addition to actions that lower the cost of building or land over time, property-based incentives also include a wide range of tax abatements or those that streamline the development process.

Common strategies include property tax assessment freezes, reductions, and exemptions. Each of these approaches effectively reduces the cost of the land by adjusting the tax an owner pays on the property with the explicit purpose of generating developer interest. Local governments usually apply property-based tax abatements in targeted redevelopment areas under conditions where developer interest is lacking. The city justifies the abatement by arguing that the incentive will generate major improvements and result in a significant increase in the assessed value of the property.

For example, Memphis' Downtown PILOT Program partially freezes the assessed value of a property at the pre-redevelopment rate for a set number of years in order to stimulate rehabilitation of blighted and historic properties (Morris 2013). From the city's perspective, the incentive attracts new development downtown and therefore increases property values and sales tax revenues even if it forgoes revenues on some part of the property's value. For instance, a ten-year freeze on a property valued at $2.5 million could save the developer $300,000 in taxes. Assuming the developer can secure adequate financing, she or he can apply this saving toward rehabilitating the structure and façade of the property. These improvements in turn increase the value of the property and mean higher future tax income for the city. Property tax exemptions are also often included in larger economic development programs such as Tax Increment Financing, which we discuss in detail below.

In addition to tax abatements, localities may also pursue land assembly, offer publicly owned land or buildings at a reduced cost, provide low-interest loans, and offer specialized grants such as façade restoration and improvement programs to spur development. Each of these strategies was instrumental in the development of the large-scale redevelopment projects discussed in the prior section. Assembling small parcels is particularly crucial. This is because many downtown blocks are comprised of smaller parcels and this approach removes one hurdle in the development process. Cities also use these incentives to encourage land-use changes. For example, many cities have targeted their older downtown office and industrial spaces for residential development. To encourage this transition, they may offer publicly owned land or buildings at a reduced cost or provide low-interest loans. Similarly, façade improvement programs, which pay for rehabilitation of historic building frontages, improve the look and value of properties. They are targeted at areas to generate new retail and business development interest.

These incentive strategies require that municipalities either invest directly or forgo future tax revenues, resulting in costs to the city. While cities use property-based incentives because they believe the incentives attract property development that they would not have otherwise, research assessing the benefits and costs of these incentive programs has found that they do not have a major impact on location decisions, for a number of reasons (Kenyon, Langley, and Paquin 2012). First, because in many states property taxes comprise a very small part of business costs, sometimes as low as 1%, reducing property taxes is insignificant for the property owner. Second, firms and developers also receive tax abatements even if they already planned to come to a location, and therefore cities often forgo revenues unnecessarily. Finally, since many places offer these abatements, developers can obtain them in most places they intend to develop. Thus the abatement in and of itself may not increase the incentive to choose one location over another.

ZONING STRATEGIES

In conjunction with offering incentives, local governments also use flexible and specialized zoning and development standards to make an area more attractive to developers or tenants. One common approach is incentive zoning. Incentive zoning allows developers to bypass or exceed established conditions usually pertaining to building height or mass. For example, to generate development in a targeted area, a city may allow a project to exceed density limits, providing it with a "density bonus," which allows increased floor space or building height. In exchange, cities will require that the recipient provide some type of public benefit at the site such as a public plaza or park.

Another common strategy is overlay zoning. This allows special provisions to be overlaid onto existing zoning in order to meet defined objectives in a specific area. The provisions can be implemented to meet a wide range of objectives. They can preserve fragile ecosystems, natural resources, historic districts, or even an area's rural character. They can be used to facilitate walkable communities, retain the urban character through urban design standards, or incentivize higher densities for transit-oriented development.

Changing parking requirements also impacts development costs. Typically, cities possess minimum parking requirements. In dense downtown areas, providing parking spaces uses valuable space that could be used in other ways, and it can add substantial costs to rents or condo prices. It can also be a challenge when adapting historic warehouses to residential uses which can require as many as two parking spots per unit for buildings that were designed without parking. Given that urban residents choose downtown locations to drive less, developers must sometimes dedicate more space to parking than actual demand necessitates. Donald Shoup (2005) has shown that providing "free" parking comes with substantial costs to land development and reduces the costs of driving, therefore encouraging more driving than people otherwise might do. In response, to encourage redevelopment, some cities now set *maximum* parking standards for new development projects, while cities such as Austin and San Francisco have removed parking minimums in their CBDs altogether.

A good example that combines these tools is Los Angeles' Adaptive Reuse Ordinance (ARO). Initiated in 1999, the ordinance offers property tax credits in exchange for ongoing preservation of historic downtown properties. The property tax credits allow owners to claim 20% of the amount they spend on rehabilitating a qualifying historic structure. In conjunction, it provides a streamlined permitting process, flexibility in fire and safety measures, does not require the provision of new parking spaces, and waives any restrictions on the number of units developed within the structure. The ordinance was initially targeted to encourage renovation of the many older, dilapidated downtown office buildings for residential use.

The ARO has been a huge success and is widely attributed with catalyzing residential growth. Since application in 1999, downtown Los Angeles has gained nearly 30,000 new residents and just under 18,000 new residential units (Guzmán 2012). In 2003, the program was expanded to other parts of the city.

Cities also attempt to create central city development by literally creating new land for development or reclaiming underutilized infrastructure. For example, New York City and developer Olympia and York built the 92-acre Battery Park City in Lower Manhattan using landfill excavated from the original World Trade Center site and sand dredged from New York Harbor. Austin, Dallas, and Fort Worth, TX have all built infrastructure to protect their downtown floodplains and opened up additional space for new development in the process. Perhaps the most high-profile example is the High Line Park in New York City (Box 6.1, Figure 6.4).

BOX 6.1 HIGH LINE PARK, NEW YORK CITY: PUBLIC SPACE OR "DISNEY WORLD ON THE HUDSON"?

New York's High Line Park has been widely lauded as a creative use of outmoded industrial infrastructure and a catalyst for urban revitalization (Birnbaum 2012, McGeehan 2011). Spearheaded by residents and developed in partnership with the Department of Parks and Recreation, the park has transformed nearly 1.5 miles of abandoned freight rail line elevated above West Manhattan into one of the city's most popular public spaces. Tourists and locals alike flock to the High Line to take in views of the city, picnic, people watch, and stroll along the old tracks landscaped with tall grasses and wildflowers. When the High Line opened in 2009, it drew an estimated 2 million visitors. In contrast to most tourist destinations, particularly those in New York, about half of the visitors are from the city. The High Line is a triumph of collaborative urban revitalization and everything a true public space should be, or is it?

The project came together in 1999 when two neighborhood residents began the Friends of the High Line to halt the planned demolition of the historic rail line. Initially, the Friends battled neighborhood property owners and city officials to see their vision of transforming the industrial ruin into a public park. After tireless efforts, they garnered the support of celebrities such as Diane von Furstenberg who built her corporate headquarters in the area, as well as Senator Hillary Clinton, Mayor Bloomberg, and the City's planning department. Each of these supporters understood the value of public green space. The City got involved following the release of an economic feasibility study by the Friends of the High Line, which showed that the project would generate $262 million in tax revenues over a 20-year period, nearly double the $150 million cost. With City support, opposing property owners withdrew their complaints and the City took ownership of the rail line (Friends of the High Line 2005). Construction began in 2006 and the first segment opened in 2009. In September 2014, the third and final segment of the High Line opened.

Although the park is owned by the City of New York and under the jurisdiction of the Department of Parks and Recreation, the Friends of the High Line raised 90% of the park's operating budget and continue to fund the majority of operations. They rely heavily on private and corporate sponsors and volunteers to keep the public space open.

Proponents boast that the High Line has brought tremendous social, economic, and environmental benefits to the city through a creative combination of historic preservation and urban design (Birnbaum 2012, McGeehan 2011). The High Line offers a popular destination that stimulates public life around a green infrastructure project. It has also become one of the city's most popular destinations, attracting 3.7 million people annually. Alongside this, the High Line has generated an estimated $2 billion in private investment in the surrounding area, 8,000 temporary construction jobs, and about 12,000 permanent jobs. New luxury apartments and hundreds of art galleries, restaurants, and boutiques have sprouted up in the surrounding area. All of this activity has boosted area property values. In one building, the price of apartments has doubled since the opening of the High Line (McGeehan 2011).

This success has inspired similar projects. Dallas has recently opened a new "deck" park, which spans a recessed freeway to link the downtown Arts District with the upscale Uptown District. Chicago, London, Philadelphia, Toronto, and Washington, DC all have plans for their own elevated parks as well (Green 2012).

While the High Line is held up as a revitalization success story, it is not without its detractors. Critics call the High Line a "tourist-clogged catwalk and a catalyst for some of the most rapid gentrification in the city's history" (Moss 2012). While creating a new and attractive public space, the High Line has simultaneously been taken over by tourists and upscale shoppers who "jam together like spawning salmon" (Moss 2012). As a result, contrary to claims of revitalization, some see the High Line as instrumental in the destruction and gentrification of its surrounding neighborhoods. They argue that the Bloomberg administration specifically supported the High Line as a tool to gentrify Manhattan's West Side in conjunction with rezoning the area for upscale apartments in 2005. This, along with the influx of art galleries into surrounding Chelsea, has pushed out working-class residents, auto-related businesses, light industry, and the retail that served them.

Others counter that such changes are an inevitable part of New York's transformation as a global city. The High Line, they say, stands as a symbol of contemporary redevelopment where the market dictates how neighborhoods develop and leaves no room for preservation of the old. To be sure, the High Line represents the ongoing challenges of revitalization in a market economy. But can one unique park really spawn the gentrification of an entire city neighborhood? Or, given the Manhattan location, would the area have gone upscale so rapidly even had the rail line been torn down? Did the City develop a public space that balances the needs of existing residents and businesses while accommodating tourism and the changing economy, or could they have done more to merge working class and creative class?

Figure 6.4 The High Line, New York City
Photograph by Carl Grodach

BUSINESS IMPROVEMENT DISTRICTS

Business Improvement Districts (BIDs) are one of the most common forms of public–private partnership engaged in urban revitalization today (Hoyt and Gopal-Agge 2007). BIDs are private organizations established to manage a neighborhood or district. They are enabled through state legislation and authorized by municipal government. The municipality determines the BIDs' powers and boundaries, but a majority of property owners must agree to participate for a BID to be established. Once a BID is enacted, all properties are assessed a fee that can be used for additional services or local improvements in the BID. While BIDs are popular for commercial property owners, they also remain a point of controversy because they do much more than simply "fill gaps in municipal services or compensate for what [local governments] do poorly" as one BID representative maintains (Levy 2001: 128). Rather, they play an important role in determining the "symbolic dimension of what the city is and for whom it is made" (Marquardt and Füller 2012).

The first BID was established in Toronto, Canada in 1969. This was followed in 1974 by the New Orleans Downtown Development District in the US. However, it was not until the 1980s and 1990s that BIDs took off as a global downtown revitalization initiative. By the end of the 1990s there were over 400 BIDs in the US, and by the mid-2000s there were 347 in Canada, 261 in Japan, 225 in Europe, and 185 in Australia (Hoyt and Gopal-Agge 2007, Mitchell 1999). According to one source, there are over 1,400 in the US today (Leinberger 2005) with nearly 70 BIDs in New York City alone (New York City Small Business Services 2014). Today, BIDs operate around the world under a variety of labels including Special Service Areas, Public Improvement Districts, and Special Improvement Districts.

Business and/or property owners within the BID zone pay annual fees to fund the district. BIDs' annual revenues can range from a few hundred to several million dollars and are frequently supplemented by other funding sources including voluntary donations and government grants (Hoyt and Gopal-Agge 2007, Levy 2001, Mitchell 1999). These fees pay for improved local services and improvements. Virtually all BIDs pay for enhanced service provision, including garbage collection and private security (Hoyt and Gopal-Agge 2007). They shape the character of the area by overseeing streetscaping, landscaping, and other capital improvement projects, and even determine the placement of benches and street lighting and dictate regulations on building façades (Ward 2006). BIDs are also heavily involved in area marketing and organize public events and festivals to promote their districts. They conduct research on housing and demographic trends, promote housing and retail development and, although they rarely initiate major redevelopment projects, they may work to broker deals. In some large cities, BIDs even provide services for homeless residents and job training (Mitchell 1999).

The rise and spread of BIDs was a response to a combination of factors we have discussed that coalesced in the 1980s including declining center city prospects, federal fiscal austerity, and the increased pressures for privatization and urban competitiveness particularly with suburban municipalities (Morçöl et al. 2008). BIDs represent a market-oriented solution to the challenge of providing quality urban services under fiscal austerity. As such, they can be considered a neoliberal, entrepreneurial urban strategy based on the belief that the private sector should take a lead role in urban revitalization and economic development (Brenner and Theodore 2002, Peyroux, Pütz, and Glasze 2012, Ward 2006). Indeed, according to a widely cited BID survey, the majority of BID directors consider their primary role as entrepreneur rather than public supervisor (Mitchell 1999).

While BIDs are a form of public–private partnership, private actors take a leadership role in service delivery, management, and governance in BID zones. As a result, BIDs are at the heart of long-standing debates over privatization, public accountability, and urban public space. Proponents argue that BIDs are a necessary actor in the revitalization of downtowns because local government does not or cannot provide necessary services. In this regard, BIDs benefit their communities by providing important consumer and public services that area businesses and residents would otherwise do without. Moreover, due to their autonomy, BIDs may be able to respond more rapidly to local issues and apply more flexible solutions than local government (Hoyt and Gopal-Agge 2007, Levy 2001).

At the same time, they control the street, urban design guidelines and public security in ways that make some people welcome and exclude others. Advocates of BIDs acknowledge that they are run by and for private interests and that they are "customer focused" (Levy 2001). The uniformed BID security patrols are "safety ambassadors who offer a friendly face on the street" (Leinberger 2005: 11) for desired customers, while undesirable street users are formally and informally excluded from the streets in BIDs (Loukaitou-Sideris and Ehrenfeucht 2009). There have been documented accounts in which BID security forces actively force homeless individuals and others perceived as a nuisance to leave the area, resulting in harassment law suits against BIDs (Batchis 2010, NCH n.d). BIDs have also been criticized for creating generic streetscapes and a mall-like atmosphere in an effort to create shopping destinations that appeal to suburban tastes (Ward 2006). In Box 9.2, we discuss how one BID has tried to address these criticisms.

Creating local districts like BIDs can have unintended consequences. While few dispute that business owners, property owners, residents, and visitors in the district benefit from their services, their actions do not reduce homelessness and other urban problems, but instead displace them to adjacent areas or other parts of the city. Instead of working towards adequate urban services for all urban neighborhoods, the localized public improvements only impact the area in the BID boundaries, and therefore they do not assist in the overall revitalization of the urban core. This ultimately contributes to spatial inequality in service delivery (Hoyt and Gopal-Agge 2007).

BOX 6.2 CAN BIDS SERVE DIVERSE RESIDENTS?
THE CASE OF SEATTLE'S METROPOLITAN
IMPROVEMENT DISTRICT

Established in 1999, Seattle's Metropolitan Improvement District (MID) provides maintenance, security, and hospitality services along with marketing and research analysis for 285 blocks in the Belltown, Denny Triangle, Pioneer Square, Retail Core, Waterfront, and West Edge neighborhoods in Seattle's downtown. With a budget of $5.6 million and over 70 MID Ambassadors who clean and patrol the streets by bike and by foot, the MID was recognized with a Merit Award in Downtown Leadership & Management by the International Downtown Association (IDA n.d., Metropolitan Improvement District n.d). Approximately half the Ambassadors are the Clean Team and half are Safety Ambassadors, the eyes and ears on the street for the police and business community. The MID also engages in other public space activities such as organizing food trucks and providing park amenities and information to visitors. The MID covers approximately 62% of the downtown core properties (Metropolitan Improvement District n.d., Downtown Seattle 2013).

Seattle's downtown neighborhoods had experienced disinvestment prior to the time they began gentrifying in the 1990s. Many people who were homeless spent time and accessed services in Pioneer Square, Belltown, and the Denny Triangle. As Seattle began to see an urban resurgence, many business leaders and other downtown stakeholders saw people who appeared homeless as a problem rather than as potential stakeholders. As with other cities, Seattle had battles over livability that sought to exclude the homeless, services for low-income residents, and permanently supportive housing from the gentrifying areas. Seattle was one of many cities that enacted ordinances prohibiting sleeping in parks and on sidewalks, panhandling, sitting on sidewalks, and loitering, and created ordinances that targeted people who appeared homeless (Gibson 2004, Loukaitou-Sideris and Ehrenfeucht 2009).

Such ordinances have been ineffective and many have been found unconstitutional. People must sleep and sit somewhere, and these ordinances can make life harder for peoplewho are homeless, but they do nothing to reduce homelessness (Ehrenfeucht and Loukaitou-Sideris 2014). Even with these ordinances, Seattle's downtown neighborhoods continued to be a space where people with mental health problems and people who appeared homeless spent time.

Realizing that the ordinances were not a solution, Seattle's MID took a different tactic and approached its job as a way to connect people with services rather than drive them from the area. MID Ambassadors provide outreach for people who are homeless, offering assistance to access shelter, food, clothing, and medical care. According to the MID, between September 1 and October 1, 2014 they assisted 508 homeless persons in finding shelter, food, clothing, and medical attention. They also wake up public sleepers in the morning, which reduces the number of trespassing calls to the police. Through its Second Change program, people who are formally homeless have had the opportunity to become MID Ambassadors. The MID also works jointly with the Union Gospel Mission to fund a mental health professional, and the Union Gospel Mission has trained the Ambassadors to both watch and respond to signs of mental illness. As people walking the streets daily, the Ambassadors regularly get to know people who are chronically homeless or who spend much time on the streets or in Occidental Park or Westlake Park. With these skills, the Ambassadors can check in on longtime residents and even de-escalate crisis situations (Burkhalter 2013, Metropolitan Improvement District n.d).

While some consider this a welcome shift from the tendency of BIDs to drive residents who are homeless from an area, others think that it is just a friendlier way to make the area safe for shoppers. In July and August 2014, MID Ambassadors enforced the "sit/lie", or pedestrian interference law, over 3,000 times, which greatly exceeds the number of times that the Ambassadors gave other types of assistance (Metropolitan Improvement District n.d). In nearby Capitol Hill neighborhood, business owners are concerned that the MID is shifting social problems to other neighborhoods (Cohen 2014). BIDs operate in a public–private realm and therefore their public duties are as important as their private successes. Can BIDs respond to the diverse and conflicting needs of different stakeholders? Or by design, will they privilege some stakeholders over others?

TAX INCREMENT FINANCING

Tax Increment Financing (TIF) is a public financing mechanism that local governments use to encourage private sector investment in an area by redirecting property tax revenue back into a designated district. When a TIF district is established, property values are assessed in a baseline year. As the area revitalizes and property values increase, the accompanying property tax increases are reinvested in the TIF district. This continues for the life of the TIF district, usually a period of 20–30 years. Like a BID, a TIF is a district-based tool. While BIDs establish new fees to pay for new services, TIFs work by reinvesting property tax revenues into the district. The tax revenues otherwise would go into the city's general fund and be used for any city expenses.

TIFs are one of the most widely applied urban revitalization tools in the US (Briffault 2010, Youngman 2011). Although initially conceived in California and Minnesota in the 1950s, the program did not experience wide application until the late 1970s and 1980s (Weber and O'Neill-Kohl 2013, Arvidson, Hissong, and Cole 2001). For instance, although Illinois embraced TIFs relatively early on in 1977, by 1984 only 26 TIF districts existed in the state. By 1990 Illinois contained 243 districts (Weber and O'Neill-Kohl 2013). Today, every state in the US with the exception of Arizona contains legislation enabling TIF districts.[1] Many cities heavily rely on TIFs. In 2001, California TIF districts took in 10% of the state's property tax revenues. In 2009 alone, Chicago garnered $1 billion in TIF funds and TIF districts encompassed 25% of the city's land (Youngman 2011).

Cities often create TIF districts to subsidize large-scale redevelopment including shopping center and historic building rehabs, new transit facilities, office complexes, sports arenas, museums, big-box stores, and upscale housing and commercial developments (Weber and Goddeeris 2007). Cities use TIF money to provide funding for new infrastructure, landscaping, streetscaping, site assembly and preparation, environmental remediation, and building renovations in commercial projects. In some instances, TIFs have been used to support specific economic development targets such as the Creative Economy TIF District in Portland, Maine (Grodach 2011), or the Kinzie Industrial Corridor TIF in Chicago (Fitzgerald and Green Leigh 2002). Although rare, TIF districts can support affordable housing and job training as well. For example, in California TIFs were required to set aside 20% of funds for "low and moderate income" housing.

TIF is structured to allow local governments to subsidize the up-front costs of redevelopment based on the future tax revenue growth. Local governments can fund TIF district improvements through a process called front-funding. In this case, municipalities offer bonds, which they pay down through the tax increment revenues. Another option, "pay as you go," requires developers to front the project costs and receive annual payments from the municipal tax increment revenues.

TIF has been popular because, in the words of an Illinois State Senator, it is a "self-help tool," a means for "local government, by its own bootstraps, cleaning up its own areas, without a real cost to the taxpayer" (Senate Transcript, 1976, quoted in Weber and O'Neill-Kohl 2013). In this sense, proponents consider TIF districts to be self-financing because they do not require additional taxes to leverage new investments. By covering major costs of development, TIF incentives help to increase the demand for hard-to-develop property where developers may not otherwise invest. Initial TIF investments lead to an increase in property values, which further attracts commercial development that replaces high-poverty areas and hires local residents who spend money in the area, thereby enhancing local sales tax revenues.

The driving purpose of the TIF strategy is to revitalize blighted areas and, with a few exceptions, all TIF districts must demonstrate that blighted conditions exist prior to establishment. The problem lies in how states define blight. As we saw in the discussion of urban renewal in chapter 3, blight continues to be defined as conditions that impede property development and appreciation. Typically, state legislatures define blight based on physical conditions and the idea that the area is not realizing the "highest and best use" of the property. Areas are considered blighted if they contain a large number of older buildings, property values below the city median, property value depreciation, slow property appreciation, structures that do not meet code standards, and/or above-average vacancy rates (Weber and Goddeeris 2007). As a redevelopment attorney claimed, "States' definitions of blight are so broad and vague that they could apply to practically every neighborhood in the country. ('Blight' can include such things as a home not having two full bathrooms or three full bedrooms)" (Youngman 2011: 322). Some states, such as Texas, do not even possess a blight requirement or require that targeted land should have any pre-existing development so long as it serves other legislative objectives (Arvidson, Hissong, and Cole 2001). In this case, TIF becomes a tool to subsidize greenfield development and can thus indirectly impair chances of revitalization in the urban core.

TIF requirements emphasize physical conditions and its success relies on property appreciation. As a result, it may not be the best strategy to improve conditions in poor communities. Like BIDs, TIFs are a market-based redevelopment strategy that gained popularity as cities faced declining funds (Weber 2002). However, the most disinvested neighborhoods are also often the most difficult to redevelop. If little or no property appreciation occurs, there will be little revenue to reinvest in the district. As a result, TIFs tend to be located in those areas with the highest likelihood for property appreciation, usually those areas near already thriving parts of the CBD. According to one study cited in Weber and Goddeeris (2007), TIF districts tend to be established in disadvantaged areas, but not the most distressed areas in a city. Indeed, even if a TIF helps an impoverished area to stabilize or to slow decline, if property values do not rise, the TIF cannot function. Another issue that critics point to is the abuse of TIF when it becomes a tool to develop bike-share programs, luxury golf courses, and other "elite amenities" to attract young urban professionals (Renn 2013, Weber and Goddeeris 2007).

A different challenge faces municipalities when deciding whether to implement TIF districts. Given that TIF districts are often established in areas with some growth potential, it is difficult to determine if the TIF is the cause of growth. Only about 20 states require that TIF districts demonstrate the new development would not have occurred "but for" the presence of the TIF. In other words, if the city is to give up tax revenues, it must be sure the TIF is the cause of redevelopment. However, not only is it very difficult to establish causation, but as with the blight requirement, even states that apply the "but for" clause may do so only superficially (Youngman 2011). As a result, while property values may increase in a TIF district it is difficult to pin down if this is due to the TIF, to a strategic location adjacent to another rapidly developing area, or if the area was attracting developer interest and poised for growth prior to TIF designation.

Even when TIF districts are successful, there are other challenges. TIFs may incentivize new development by attracting investment from other parts of the city. In other words, TIFs may shift the geography of development rather than sparking the creation of new businesses, jobs, and sales tax revenue. This is not necessarily bad. In such instances, TIFs can be applied as a means to redistribute economic development opportunities in a city or region. Finally, in many states, TIFs' role in area redevelopment comes at the expense of diverting revenue from other government services, although legislation in states such as Texas protects the tax revenue that goes to public schools.

The TIF strategy offers an opportunity for urban revitalization. By restricting TIFs to areas in need of development assistance and requiring programs to respond to community needs, whether affordable housing, underserved commercial activities, job training, or brownfield remediation, for example, they can deliver benefits for residents.

LINKAGE POLICIES

Businesses and developers have come to see incentives as a standard part of the development process. To receive the incentives, they may commit to social benefits in their development process, promise to provide a certain number of jobs, or contribute to funding public space improvements. Yet many communities have lost out because recipients of tax abatements and other incentives do not deliver the benefits they promised. Firms may fall short on the number of jobs guaranteed in a contract, provide jobs primarily to candidates outside the community, or provide only minimum-wage or temporary jobs without benefits. New projects may create an undue burden on infrastructure or increase traffic congestion and air pollution for residents. Additionally, successful projects may actually spark an increase in the value of surrounding property to the extent that it prices out existing residents and small businesses.

To ensure that local communities reap benefits from the new development, some cities incorporate linkage policies into redevelopment contracts. Linkage policies require large-scale property developers to provide public benefits in exchange for tax abatements and other incentives. These may include funding for on- and off-site affordable housing, workforce development, or local hiring requirements. They may also involve the provision of public goods including public art, park maintenance, or historic preservation. Typically, recipients make a contribution based on an agreed-upon percentage of the property square footage over a certain minimum. Linkages may apply to hotel, retail, entertainment, and office property.

First source hiring agreements are one of the most common linkage policies. In this case, firms or developers that accept government assistance contractually agree to first consider a set number of local workers for open positions before looking outside the community. While initially used in firm-based incentives, they are increasingly also a part of large land development deals where a developer accepts benefits from local government in exchange for developing a property under the argument that the development should benefit the surrounding community. First source hiring agreements will often target residents in the surrounding area, individuals who lost their jobs due to the new development, or low-income or special needs residents.

Cities like Boston go a step further and also require developers of new commercial projects receiving a zoning variance such as expanded density allowances to contribute to a jobs training fund (Barnes n.d). The Neighborhood Jobs Trust program applies to projects over 100,000 square feet and requires developers to pay the linkage fee, based on square footage, to the general fund or toward training specifically related to the project. The Trust funds job training services, employment counseling and placement services, and adult education services for low- and moderate-income residents.

While San Francisco operates a housing linkage policy, the City also negotiates on a project-by-project basis. In major redevelopment projects like Yerba Buena Center (YBC), the San Francisco Redevelopment Agency began in the 1980s to negotiate with developers for first source hiring contracts, affordable housing, land set asides, and ground floor space for cultural institutions in new office projects. Further, the Agency requires that major YBC commercial projects contribute to a fund to support a portion of the operation and maintenance costs for cultural organizations in the redevelopment area (Grodach 2010).

The YBC case reflects an example of what are now more commonly known as Community Benefits Agreements (CBA). CBAs differ, however, in that they tend to stem from community-based organizations whereby developers agree to meet a set of conditions in exchange for project support from community organizations and labor unions. Community benefits commonly include first source hiring, living wage agreements, and funding for affordable housing or open space. In exchange, developers gain assurance of a smoother development process (Wolf-Powers 2010). One of the first CBAs was negotiated in 2001 for the $2.5 billion, 27-acre LA Live sports and entertainment district in Los Angeles. This CBA resulted in interest-free loans from the developer for the production of new affordable housing units, funding for a local hiring and training program, a hiring obligation of up to 50% from CBA zip codes, and 300 full and 1,700 part-time living wage jobs (Saito and Truong 2014).

Today, most cities try to avert the negative consequences of development by carefully awarding incentives and including "clawback" clauses in the contract to guarantee benefits. Clawbacks are a contractual obligation in which the incentive recipient agrees to provide a certain number of jobs or hire a predetermined number of people from the local area, for instance. If they fail to meet the agreement, the clawback allows the city to recapture the lost money.

While not altering the entrepreneurial approach to redevelopment, linkage policies change the development context to guarantee a more equitable distribution of benefits for those affected by redevelopment. However, linkage fees are most successful when cities anticipate strong demand for property and business development. In struggling downtowns and slow-growth cities, they may actually impede redevelopment prospects. Moreover, the project-by-project nature of CBAs is costly and time-consuming for both parties.

SUMMARY

The downtown urban core in many cities has emerged as a center of entertainment and consumption. Large-scale projects and mixed-use districts comprised of arts, sports, shopping, and residential development define the new downtown. These transformations are due to the heightened competition between cities for private investment. They have also come about through changes in the downtown power structure, which has focused on luring new residents and visitors away from the suburbs and appealing to young urban professionals and empty nesters by providing a safe and fun urban experience. Cities have incentivized large-scale redevelopment through an array of strategies that boost demand by lowering development costs through tax abatements, special financing, and development zones as well as through zoning and historic preservation strategies that encourage new uses of old buildings. These activities have sparked major improvements to the look and feel of places and have dramatically raised property values, but do not always catalyze wider social and economic benefits. Although linkage policies and community benefits agreements help to ameliorate this situation, social and economic equity remain the biggest challenges for the continued revitalization of downtown and the urban core.

STUDY QUESTIONS

1   Why do downtown redevelopment strategies concentrate on consumption-based activities?
2   What appear to be the most effective downtown revitalization strategies? Why? How would you adapt them to work in the economic and political context of your city?
3   Do you feel that the privatization of public space is an issue that city officials should be concerned with when incentivizing new urban revitalization projects?

NOTE

1   In 2011 California eliminated all of its redevelopment agencies, which meant an end to TIF and the transfer of their funds to Redevelopment Trust Funds, which are managed in the interim by "successor agencies" (Los Angeles County n.d.).

REFERENCES

Arvidson, Enid, Rod Hissong, and Richard Cole. 2001. "Tax increment financing in Texas: survey and assessment." In *Tax increment financing and economic development: uses, structures, and impact*, edited by Craig L. Johnson and Joyce Y. Man. Albany: State University of New York Press.

Barnes, Ken. n.d. *Guide to the Neighborhood Jobs Trust*. Boston: City of Boston.

Batchis, Wayne. 2010. "Business improvement districts and the constitution: the troubling necessity of privatized government for urban revitalization." *Hastings Constitutional Law Quarterly* 38(1): 91–130.

Birnbaum, Charles. 2012. "The Real High Line effect – a transformational triumph of preservation and design." *Huffington Post* June 19. www.huffingtonpost.com/charles-a-birnbaum/ the-real-high-line-effect_b_1604217.html

Brenner, Neil and Nik Theodore. 2002. "Cities and the geographies of 'actually existing neoliberalism'." *Antipode* 34(3): 349–379. doi: 10.1111/1467-8330.00246

Briffault, Richard. 2010. "The most popular tool: tax increment financing and the political economy of local government." *University of Chicago Law Review*: 77(1): 65–95.

Burkhalter, Aaron. 2013. "A band-aid for breakdowns." *Real Change* May 15. http://realchangenews.org/index.php/site/archives/7805

Chapple, Karen, Shannon Jackson, and Anne J Martin. 2010. "Concentrating creativity: the planning of formal and informal arts districts." *City, Culture and Society* 1(4): 225–234. doi: 10.1016/j.ccs.2011.01.007

Clark, Terry Nichols. 2004. *The city as an entertainment machine*. Vol. 9. Amsterdam: Elsevier/JAI.

Cohen, Bryan. 2014. "Another view of Capitol Hill homeless stats shows shift from downtown." *Capitol Hill Seattle Blog* May 19. www.capitolhillseattle.com/2014/05/another-view-of-capitol-hill-homeless-stats-shows-shift-from-downtown/

Dávila, Arlene M. 2004. *Barrio dreams: Puerto Ricans, Latinos, and the neoliberal city*. Berkeley: University of California Press.

Downtown Seattle. 2013. "The Seattle City Council voted today to renew the Metropolitan Improvement District (MID) in Downtown Seattle for another 10 years." downtownseattle.com May 6. www.downtownseattle.com/blog/2013/05/06/seattle-city-council-approves-mid-renewal

Ehrenfeucht, Renia and Anastasia Loukaitou-Sideris. 2014. "The irreconcilable tension between dwelling in public and the regulatory state." In *The informal American city: beyond day labor and taco trucks,* edited by Vinit Mukhija and Anastasia Loukaitou-Sideris. Cambridge, MA: MIT Press.

Fainstein, Susan S. 2005. "The return of urban renewal: Dan Doctoroff's grand plans for New York City." *Harvard Design Magazine* Spring/Summer (22): 1–5.

Fitzgerald, Joan and Nancey Green Leigh. 2002. "Job-centered economic development: an approach for linking workforce and local economic development." In *Economic revitalization: cases and strategies for city and suburbs*, edited by Joan Fitgerald and Nancey Green Leigh. Thousand Oaks, CA: Sage.

Frieden, Bernard J and Lynne B Sagalyn. 1991. *Downtown, inc: how America rebuilds cities*. Cambridge, MA: MIT Press.

Friends of the High Line. 2005. "Major federal authorization for High Line project." *Newsletter* 13 June. http://assets.thehighline.org/original_site/newsletters/061305_pr.html

Gibson, Timothy A. 2004. *Securing the spectacular city: the politics of revitalization and homelessness in downtown Seattle*. Lanham, MD: Lexington Books.

Glaeser, Edward L, Jed Kolko, and Albert Saiz. 2001. "Consumer city." *Journal of Economic Geography* 1(1): 27–50.

Green, Jared. 2012. "Everybody wants a High Line." *The Dirt* 26 July. http://dirt.asla.org/2012/07/26/everybody-wants-a-high-line/

Grodach, Carl. 2010. "Beyond Bilbao: rethinking flagship cultural development and planning in three California cities." *Journal of Planning Education and Research* 29(3): 353–366. doi: 10.1177/0739456X09354452

Grodach, Carl. 2011. "Cultural institutions: the role of urban design." In *Companion to urban design*, edited by Tridib Banerjee and Anastasia Loukaitou-Sideris. New York: Routledge.

Guzmán, Richard. 2012. "Adapting the Adaptive Reuse Ordinance." *Los Angeles Downtown News* 31 January. www.ladowntownnews.com/news/adapting-the-adaptive-reuse-ordinance/article_f9bf41da-493a-11e1-b6c4-0019bb2963f4.html

Hackworth, Jason and Neil Smith. 2001. "The changing state of gentrification." *Tijdschrift voor economische en sociale geografie* 92(4): 464–477.

Hammett, Jerilou and Kingsley Hammett. 2008. *Suburbanization of New York*. New York: Princeton Architectural Press.

Hannigan, John. 1998. *Fantasy city: pleasure and profit in the postmodern metropolis*. London: Routledge.

Hoyt, Lorlene and Devika Gopal-Agge. 2007. "The business improvement district model: a balanced review of contemporary debates." *Geography Compass* 1(4): 946–958.

IDA. n.d. "IDA downtown achievement awards." Washington, DC: International Downtown Association. www.ida-downtown.org/eweb/dynamicpage.aspx?webcode=2014winners

Judd, Dennis R. 1999. "Constructing the tourist bubble." In *The tourist city*, edited by Dennis R Judd and Susan S Fainstein. New Haven, CT: Yale University Press.

Judd, Dennis R and Susan S Fainstein. 1999. *The tourist city*. New Haven, CT: Yale University Press.

Kenyon, Daphne A, Adam H Langley, and Bethany P Paquin. 2012. *Rethinking property tax incentives for business*. Cambridge, MA: Lincoln Institute of Land Policy.

Lang, Robert E and Jennifer LeFurgy. 2003. "Edgeless cities: examining the noncentered metropolis." *Housing Policy Debate* 14(3): 427–460. doi: 10.1080/10511482.2003.9521482

Leinberger, Christopher B. 2005. *Turning around downtown: twelve steps to revitalization*. Washington, DC: Brookings Institution Center on Urban and Metropolitan Policy.

Leinberger, Christopher B. 2009. *The option of urbanism: investing in a new American dream*. Washington, DC: Island Press.

Levy, Paul R. 2001. "Paying for the public life." *Economic Development Quarterly* 15(2): 124–131.

Lloyd, Richard. 2010. *Neo-bohemia: art and commerce in the postindustrial city.* New York: Routledge.

Los Angeles County. n.d. "About redevelopment dissolution." http://redevelopmentdissolution.lacounty.gov/wps/portal/rdd/about

Loukaitou-Sideris, Anastasia, and Renia Ehrenfeucht. 2009. *Sidewalks: conflict and negotiation over public space (urban and industrial environments)*. Cambridge, MA: MIT Press.

Marquardt, Nadine and Henning Füller. 2012. "Spillover of the private city: BIDs as a pivot of social control in downtown Los Angeles." *European Urban and Regional Studies* 19(2): 153–166. doi: 10.1177/0969776411420019

McGeehan, Patrick. 2011. "The High Line isn't just a sight to see; it's also an economic dynamo." *New York Times* June 5. www.nytimes.com/2011/06/06/nyregion/with-next-phase-ready-area-around-high-line-is-flourishing.html?_r=0

Metropolitan Improvement District. n.d. www.downtownseattle.com/mid

Mitchell, Jerry. 1999. *Business improvement districts and innovative service delivery*. New York: PricewaterhouseCoopers Endowment for the Business of Government.

Morçöl, G, L Hoyt, J Meek, and U Zimmermann, eds. 2008. *Business improvement districts: research, theory, and controversies*. Boca Raton, FL: CRC Press.

Morris, Paul. 2013. "Downtown PILOT program increases local tax revenue 2013." Memphis, TN: Downtown Memphis Commission. www.downtownmemphiscommission.com/1/post/2013/09/downtown-pilot-program-increases-local-tax-revenue.html

Moss, Jeremiah. 2012. "Disney World on the Hudson*." New York Times* August 21. www.nytimes.com/2012/08/22/opinion/in-the-shadows-of-the-high-line.html

NCH. n.d. "A dream denied: The criminalization of homelessness in U.S cities." Washington, DC: National Coalition for the Homeless. www.nationalhomeless.org/publications/crimreport/casesummaries_1b.html.

New York City Small Business Services. 2014. *Neighborhood development*. City of New York. www.nyc.gov/html/sbs/html/neighborhood_development/bids.shtml

Peyroux, Elisabeth, Robert Pütz, and Georg Glasze. 2012. "Business Improvement Districts (BIDs): the internationalization and contextualization of a 'travelling concept'." *European Urban and Regional Studies* 19(2): 111–120. doi: 10.1177/0969776411420788

Positively Cleveland. 2014. "You're welcome." www.thisiscleveland.com

Renn, Aaron M. 2013. "Well-heeled in the Windy City: Rahm Emanuel splurges on amenities for the elite, while poor and middle-class Chicagoans suffer." *City Journal* October 16.

Saito, Leland and Jonathan Truong. 2014. "The LA Live Community Benefits Agreement: evaluating the agreement results and shifting political power in the city." *Urban Affairs Review* March 25. doi: 10.1177/1078087414527064

Shoup, Donald. 2005. *The high cost of free parking*. Washington, DC: APA Planners Press.

Ward, Kevin. 2006. "'Policies in motion', urban management and state restructuring: the trans-local expansion of Business Improvement Districts." *International Journal of Urban and Regional Research* 30(1): 54–75.

Weber, Rachel. 2002. "Extracting value from the city: neoliberalism and urban redevelopment." *Antipode* 34(3): 519–540.

Weber, Rachel Nicole and Laura Goddeeris. 2007. *Tax increment financing: process and planning issues*. Cambridge, MA: Lincoln Institute of Land Policy.

Weber, Rachel and Sara O'Neill-Kohl. 2013. "The historical roots of tax increment financing, or how real estate consultants kept urban renewal alive." *Economic Development Quarterly* 27(3): 193–207. doi: 10.1177/0891242413487018

Wolf-Powers, Laura. 2010. "Community benefits agreements and local government: a review of recent evidence." *Journal of the American Planning Association* 76(2): 141–159. doi: 10.1080/01944360903490923

Youngman, Joan. 2011. "TIF at a turning point: defining debt down." *State Tax Notes* 60: 321–329.

Zukin, Sharon. 2008. "Consuming authenticity: from outposts of difference to means of exclusion." *Cultural Studies* 22(5): 724–748.

Zukin, Sharon. 2010. *Naked city: the death and life of authentic urban places*. Oxford: Oxford University Press.

7

Revitalizing neighborhoods with affordability and opportunity

Housing policies and other routes to neighborhood change

In many cities, downtowns have experienced major reinvestment and previously disinvested neighborhoods in the urban core have gentrified. In many other neighborhoods, residents continue to live in depressed areas. Some impoverished neighborhoods are isolated in the suburbs and others persist adjacent to their gentrifying urban counterparts. Public intervention through neighborhood revitalization programs is necessary because of this perpetual cycle of uneven development and the reproduction of unequal opportunity.

Attention to neighborhood revitalization is crucial because where we live and grow up can have a profound effect on our life opportunities, playing a role in everything from education and employment options to the relationships we rely on to find work. Those who grow up in neighborhoods defined by concentrated poverty—areas with a high density of low-income households—are at a disadvantage. Their neighborhoods often lack access to quality education, healthy food, good transportation, and opportunities for career employment. Discrimination in credit and housing markets, city zoning, municipal ordinances, and urban policies historically have limited access to good housing and opportunity-rich neighborhoods for low-income residents who are disproportionately people of color (Massey and Denton 1993, Wilson 1996).

In short, poor neighborhoods are underserved in multiple, interrelated ways. Many people continue to live in areas defined by concentrated poverty. Moreover, mechanisms to ensure that neighborhoods experiencing reinvestment or gentrification remain affordable for low-income residents are necessary but too often ignored. Neighborhood redevelopment, if left to market forces alone, will often result in the displacement of low-income residents rather than ensuring that they can stay and take advantage of the changes.

Determining the most effective ways to facilitate better circumstances for low-income residents and create economic opportunities that all people can take advantage of has been challenging. Addressing this challenge requires a twofold response. We need to ensure that all *neighborhoods* have high-quality housing with good urban services and amenities and ensure that all *people* have access to high-quality housing, jobs, and education.

Housing has been the prime front on which public, private, and community sector groups attempt to improve neighborhood conditions and an important route to change for low-income residents. Housing policy remains important because having secure and affordable housing can improve people's lives. Housing is also a highly visible investment that can provide an important catalyst to additional neighborhood development.

Housing alone is not sufficient, however. Some argue that we need market-based solutions to attract new businesses and jobs to revitalize struggling neighborhoods (Porter 1997), while others advocate for comprehensive community initiatives (CCI) to improve neighborhood conditions through place-based targeting across multiple fronts including housing, employment, education, transportation, and health resources.

However, the dearth of affordable housing remains a significant overriding problem in many regions. Today, in 90 US cities, median rents exceed the standard threshold for housing affordability (30% of median gross income) (Dewan 2014). In 2011, 20.6 million households spent over half of their income on housing and just 6.8 million housing units qualified as affordable to them (Joint Center for Housing Studies 2013). Moreover, the US Department of Housing and Urban Development (HUD) reports that between 2001 and 2011, the US experienced a dramatic 69% increase in "worst case needs" households (to 8.48 million), those low-income renters that do not receive federal housing assistance and spend over 50% of their income on rent (HUD 2013a).

In the following sections, we discuss different types of housing policies and how they are intended to remedy challenges facing people living in low-income neighborhoods. Following this, we provide an overview and evaluation of the most important programs for providing affordable housing and addressing concentrated poverty. Finally, we examine two other routes toward neighborhood revitalization, Comprehensive Community Initiatives (CCI), which attempt a broad-based neighborhood revitalization agenda, and Empowerment Zones, which are an effort to bring jobs to low-income, job-poor communities. While this chapter focuses largely on housing, it is worth keeping in mind that many of the policies and programs we discuss throughout the book can be implemented in different types of neighborhoods.

TYPES OF HOUSING POLICIES AND THEIR CONTRIBUTION TO NEIGHBORHOOD REVITALIZATION

Housing policy has been an essential component of revitalization efforts since the Housing Act of 1937, which initiated a formal affordable public housing program in the US. Since that time, the federal government has pursued a range of programs and approaches. Beginning in the 1930s, housing policy took on a supply-side or project-based approach. Under a supply-side approach, the federal government built housing projects to increase the supply of housing affordable to working-class and low-income households. In conjunction, it created a larger demand for single family homeownership and facilitated home construction through mortgage insurance programs that helped more households afford to buy their own house, as we discussed in chapter 2.

By the late twentieth century and into the twenty-first century, the federal government began to demolish and redevelop its aging public housing units and replace them with projects that contain a lower proportion of affordable public housing units interspersed with market rate units. Today, supply-side programs are geared primarily toward these mixed-income housing projects and are produced by private developers and community organizations often in conjunction with state and local governments. The primary project-based programs in the US include HOPE VI and its successor program Choice Neighborhoods, which provide grants to stimulate affordable housing production. Another key supply-side program is the Low Income Housing Tax Credit (LIHTC) program, which provides tax abatements to incentivize the development of affordable housing. Mixed-income housing is increasingly common and widely applied outside the US as well as in Europe, Australia, and Canada (Fraser et al. 2012).

Since the 1970s, the bulk of funding for affordable housing has turned to demand-side, or tenant-based policies, which are geared toward helping low-income households afford units already on the market. By 1997, nearly three-quarters of new federal rental assistance funds went to tenant-based subsidies while the remaining funds went toward project-based assistance programs (Goetz 2003).

Tenant-based policies primarily take the form of vouchers (Section 8 or Housing Choice Vouchers). Vouchers are intended to help recipients afford rents in the private market and to give individuals more choice in their residential location. However, they do not address the supply of affordable housing. Housing vouchers are provided to qualifying individuals including residents of older public housing projects slated for demolition or those in projects that have reverted to market rate rents. Other tenant-based examples include specialized programs that attempted to move public housing residents to lower-poverty areas such as the Gautreaux Assisted Housing Program in Chicago and the Moving to Opportunity demonstration program.

In sum, supply- and demand-side approaches address different sides of the housing question. Supply-side strategies compensate for the dearth of affordable housing units by producing more affordable units. Demand-side strategies are appropriate in markets where the overall supply is adequate, but incomes are too low for households to find affordable units.

Housing policies, like other policies intended to help low-income residents in the US, consist of a mix of people- and place-based policies. As we discussed in chapter 1, people-based policies focus on a person or household. For example, housing vouchers assist individuals and families in meeting their housing needs regardless of location. Other forms of people-based assistance include the Temporary Assistance for Needy Families program (TANF, sometimes called welfare), the Supplemental Nutrition Assistance Program (SNAP, also called food stamps), or health insurance.

Place-based initiatives focus on improving the conditions of a particular area. Place-based housing programs include the HOPE VI and Choice Neighborhoods programs that we discuss in more detail in the next section. Other place-based initiatives might include community health clinics or incentives to locate grocery stores in an underserved neighborhood.

While both people- and place-based policies can contribute to urban revitalization, they do so in different ways. People-based housing policies, in principle, give households the opportunity to choose where and how they live. They also infuse resources into the households. As family members meet their daily needs, they contribute to the demand for goods and services in the area where they live, thereby supporting retail and services. Place-based policies can improve an area by directing resources to a targeted area and attracting residents and visitors that help stimulate neighborhood revitalization. Place-based policies also recognize that resources are unevenly distributed and that people cannot always easily move. Providing money for food, for example, might be insufficient if there are no grocery stores nearby.

Housing policy today is geared toward both people-based, demand-side policies (vouchers) and place-based, mixed-income housing. People-based programs influence the demand side of housing while place-based policies tackle housing supply. While vouchers reveal little concrete evidence of moving people out of poverty and improving life conditions, mixed-income housing has had a major impact in improving conditions in high-poverty neighborhoods, but considerably less success in helping people actually move out of poverty.

HOUSING POLICY AND CONCENTRATED POVERTY

The structure of contemporary housing programs has come about in large part due to the belief that concentrated poverty has debilitating effects on individuals and communities. Since the 1960s, when the federal public housing program focused largely on the neediest families, it has contributed to concentrating very low-income households in neighborhoods with poor access to jobs, transit, and urban services. As a result, the program also perpetuated racial segregation. In 1990, single-parent households comprised nearly 43% of all public housing residents and 82% were people of color (Rosen and Dienstfrey 1998). Because funding was not adequate to cover operating and maintenance costs for many projects, US public housing quality declined. It became synonymous with very low-income black residents isolated in the most dilapidated and dangerous living environments. These conditions were further exacerbated by discriminatory siting practices (Goetz 2013, Massey and Denton 1993, Oakley et al. 2011).

Larger structural forces are equally important factors that explain the concentration of poverty, and more residents live in high-poverty neighborhoods than in public housing. The deindustrialization and economic restructuring discussed in chapter 4 resulted in African American and Latino communities living in declining urban core neighborhoods with few job prospects. Single-parent households—who primarily had access to minimum-wage jobs without affordable childcare—had no way to raise themselves out of poverty (Wilson 1987, 1996). During this time, the number of people living in high-poverty areas doubled and the physical size of concentrated poverty neighborhoods increased dramatically (Jargowsky 2003). As more and more employment opportunities moved to the suburbs and overseas, a "spatial mismatch" between people and jobs became pronounced. Discrimination intensified existing racial segregation and contributed to unequal opportunities in housing and labor markets as well as education (Massey and Denton 1993).

Concentrated poverty worsened in the late 1960s and 1970s as suburban housing markets became more open to a rising black middle class who, like the white middle class before them, fled urban neighborhoods. Low-income blacks and other people of color continued to face exclusion from many suburban areas due to zoning restrictions and opposition to the construction of apartment units and affordable housing (Joseph 2006, Wilson 1987).

Living in areas with high levels of poverty magnifies the challenges that low-income households face, and it has been an enduring problem. After substantial decline in the 1990s, the population in high-poverty neighborhoods grew by one-third during the 2000s with African American residents comprising 45% of these neighborhoods but only about 12% of the US population. Moreover, signaling the gentrification of the central city, high-poverty neighborhoods grew over twice as fast in suburban areas as in the urban core (Kneebone and Berube 2013).

Given these persistent conditions, poverty deconcentration became a priority for urban revitalization policy. Proponents of policies aimed at poverty deconcentration and mixed-income communities argue that such strategies can reduce the social and economic marginalization of low-income households by improving access to quality housing and education, employment opportunities, safer neighborhoods, and basic services (Fraser et al. 2012, Joseph 2006, Oakley et al. 2011, Chaskin and Joseph 2011). The presence of higher-income residents can improve the overall neighborhood quality of life because middle- and upper-class households' demand for higher-quality goods and services and the perception by businesses of their stronger purchasing power can attract investment and services to a neighborhood.

The argument for income mixing and poverty deconcentration nevertheless faces criticism. Although some also argue that mixed-income neighborhoods can develop bridging social capital where lower-income residents develop relationships that lead to better employment opportunities, there is no clear evidence on how or if bridging social capital develops. Simply housing the poor and wealthy together may not automatically create improved living conditions for all. This may be particularly challenging in terms of fostering social networks (Chaskin and Joseph 2011, Joseph 2006).

Another critique is that demolishing low-income neighborhoods to develop mixed-income communities boils down to state-initiated gentrification (Bridge, Butler, and Lees 2012, Chaskin and Joseph 2011). As the wealthy bring neighborhood improvements, they also boost area demand and increased rents that may price out existing residents. When a HOPE VI project breaks up the concentration of poverty in an older public housing complex, it also reduces the number of affordable units on the market and helps to redevelop the center city for higher-income groups (Fraser et al. 2012, Hackworth 2007, Jones and Popke 2010). It is possible that, rather than improving neighborhood services and quality of life for low-income residents, the arrival of higher-income residents actually prices them out of the neighborhood and displaces them into other high-poverty areas.

Many people also challenge the tendency to frame poverty as an individual issue rather than the result of structural forces. Neoliberal urban policy places emphasis on housing as an issue of personal responsibility and self-sufficiency (Hackworth 2007). For example, HOPE VI requires residents to perform community service and engage in "family self-sufficiency" programs that provide counseling and social service assistance. While these programs may be helpful for many residents, they are of little value without recognition of the larger structural barriers to upward mobility that many individuals face. People of color historically have been excluded from quality education, employment, credit, and housing markets through both formal and informal means. As a result, even if individuals work hard and develop social capital, larger social and economic conditions may override individual dispositions making it difficult for neighborhood improvement to occur through income mixing alone.

HOUSING POLICIES TO DISTRIBUTE AFFORDABLE HOUSING

Housing policies retain a prime place in the urban revitalization agenda. At the federal level, urban renewal, CDBGs, HOME grants and LIHTC have been important sources of housing funding. Since the 1980s cutbacks in federal urban funding, community-based organizations including community development bank programs and foundations have emerged as important players in affordable housing provision. At the same time, newer housing policies call for more active private sector involvement. The federal programs HOPE VI and its successor Choice Neighborhoods reflect the public–private–community partnership model as does the LIHTC program. In this section, we discuss the most common housing programs. On the demand-side, we survey Housing Choice Vouchers (HCV) and related programs; on the supply-side, we examine the outcomes of LIHTC and HOPE VI.

VOUCHERS AND MOBILITY PROGRAMS

The first voucher program in the US emerged out of Section 8 of the Housing and Community Development Act of 1974, the same housing act that ended urban renewal. The Section 8 voucher program was initiated in part due to growing criticism of public housing and urban renewal as costly and ineffective. Housing vouchers are provided to help qualifying individuals afford market rate apartments. The participants pay a portion of the rent and the voucher covers the difference between the participants' share and fair market rent. Because most housing voucher programs have long waiting lists, not all qualified households receive a voucher. Further, in many tight housing markets, landlords have less incentive to participate in the program because they can more easily find people able to pay the market rate.

Initially, vouchers covered the difference between the regional fair market rent and a percentage of the household income and could be applied in that region alone. Into the 1980s and 1990s the federal government experimented with vouchers as a means of poverty deconcentration by allowing recipients to move to areas where rents exceeded fair market rent limits. In 1998, the federal government made this a permanent feature of the voucher program and required recipients to move to low-poverty areas under the Housing Choice Voucher program (Goetz 2003). In addition, HCV provides counseling to assist households in selecting neighborhoods.

The voucher program got an early boost from demonstration projects that focused on relocating households out of high-poverty areas. One, the Gautreaux Assisted Housing Program, was the result of the *Gautreaux et al. v. Chicago Housing Authority 1969* desegregation court case. In brief, the court found that the Housing Authority concentrated poor, minority households through siting of public housing and stipulated that they must disperse future public housing and provide former residents with vouchers to move to neighborhoods that are less than 30% black (Polikoff and Page 2007). Because evaluations showed evidence of improvements for participant households, the program influenced the development of future voucher programs. Another demonstration project that encouraged the formulation of the HCV program, Moving to Opportunity (MTO), assisted residents in high-poverty areas in five cities to move to areas where the poverty rate was under 10%. However, studies show mixed outcomes. In some cities residents moved to safer, lower-poverty neighborhoods and experienced better school performance, mental health, and employment, while in others there was no observable benefit (Briggs, Popkin, and Goering 2010, McClure 2010).

In 2008, over 2.1 million households received HCVs (McClure 2010). However, demand far exceeds supply. Waiting lists are long, HUD recommends 12–24 months, and many cities close them to control extremely high demand. Beyond failure to meet demand, there is limited evidence that vouchers help move people out of high-poverty areas. Just 17% of voucher holders live in low-poverty neighborhoods. In central city areas, over one-third of HCV households live in neighborhoods with over 30% poverty (McClure 2010).

One explanation for this condition is simply the lack of affordable housing located in neighborhoods where poverty is not a concern. If HCVs were restricted to neighborhoods of 10% or less poverty, only about 5.2 million units would be available of the 19 million units that currently qualify based on fair market rent levels (McClure 2010). This problem is compounded by NIMBYism. Voucher programs allow municipalities to establish residency preferences for admission to the waiting list, which may restrict some applicants (Goetz 2003).

Alongside these challenges, there is scant evidence that voucher holders that move to low-poverty neighborhoods experience significant improvements. While some studies find that HCV recipients report improvement in overall neighborhood conditions, including lower poverty rates, other studies report that employment levels decline significantly and that school performance is unchanged (Basolo 2013).

LOW INCOME HOUSING TAX CREDITS

The LIHTC program is the largest source of funding for new, affordable rental housing units in the US. Created under the Tax Reform Act of 1986, the program provides nearly $8 billion in tax credits annually and has produced over 2.3 million housing units in 37,500 projects through 2011 (HUD n.d.-c). LIHTC seeks to encourage private sector construction of affordable housing, but not through direct subsidies like other housing programs. Instead, the program supplies tax credits to incentivize new development.

Each state has a housing agency that allocates the tax credits based on criteria set forth in the state's Qualified Allocation Plan. The state's housing agency allocates tax credits to housing developers, whether private entities or non-profit and community-based organizations, through a competitive process. The developers can use the tax credits to reduce their tax liability. In many cases, recipients transfer their credits to equity investors or investor syndicates to raise capital to fund their project. The equity investors then use the tax credits. The Internal Revenue Service allocates each state funds on a per capita basis. States with larger populations receive more LIHTC funding than states with smaller populations.

Typically, LIHTC projects receive a 9% annual tax credit for ten years, although the amount can vary by property cost and the proportion of low-income units in the proposed project. Eligible projects must contain 40% of units below 60% of the area median family income or 20% of units below 50% of the median for a minimum of 15 years. However, most projects will dedicate the majority or even all units as low-income to maximize their tax credit (Schwartz 2011). Projects located in qualified census tracts (QCT) or Difficult Development Areas (DDA) are eligible to claim 30% more in tax credits.[1]

LIHTC has created a vast amount of affordable housing and has created more units in lower-poverty neighborhoods than public housing. However, given the structure of the program, the majority of LIHTC projects are located in central cities and a significant share is in high-poverty neighborhoods. In particular, the larger tax credit provision for projects in QCTs promotes the construction of LIHTC units in high-poverty areas. In total, about 29% of LIHTC projects are in QCTs (Schwartz 2011). The potential for concentrating poverty through the program is exacerbated by the fact that, rather than requiring or incentivizing higher-income units as well, the program allows developers to maximize their credits by including more LIHTC units in the project. Conversely, another issue is that project owners may revert units to market rate following the 15 year requirement on tax credit eligibility. This may mean the reduction of affordable units over time.

Nationally, estimates place the percentage of LIHTC units in low-poverty neighborhoods in the range of 22–33%, but a similar amount (22%) are in areas where poverty rates are high (over 30%). While the track record is better than public housing (38% of units are in high-poverty areas) it exceeds the 17% of voucher holders in high-poverty neighborhoods. However, just 12.5% of LIHTC projects in suburban areas are in high-poverty areas and LIHTC contains the largest proportion of housing in areas with low poverty (Dawkins 2013, McClure 2010, Schwartz 2011).

The issue of LIHTC funding and poverty concentration is currently before the Supreme Court. In the brief, *Texas Department of Housing and Community Affairs v. the Inclusive Communities Project*, the Inclusive Communities Project in Dallas, TX claims that the Housing Department's administration of LIHTC intensifies segregation in the city. As Figure 7.1 shows, Dallas' LIHTC developments are predominantly located in census tracts with a non-white population of 80% or more. The brief argues that such "seemingly race-neutral government decisions" limit low-income minority families to living in segregated neighborhoods with fewer services than others and perpetuate historic patterns of segregation (Haas Institute 2015).

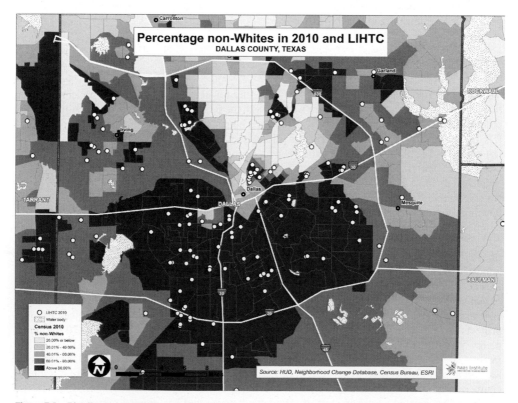

Figure 7.1   Distribution of LIHTC projects, Dallas County, TX, 2010; 72% of projects are in predominantly non-white census tracts

Image courtesy of the Haas Institute

In sum, while the LIHTC program has contributed significantly to the production of affordable housing, its role in deconcentrating poverty is in question. However, the fact is that most LIHTC tenants have incomes above the poverty level (about 30% of area median income) because the program requires 50–60% of median. Ironically, despite its name, LIHTC may actually be reducing poverty concentration because it does not provide for the households with the lowest incomes.

HOUSING OPPORTUNITIES FOR PEOPLE EVERYWHERE (HOPE VI)

Although LIHTC and the HCV programs have received considerably more funding, the HOPE VI program has been the most controversial and highest-profile of the three. Initiated in 1992, HOPE VI issued over $6 billion in grants to local housing authorities. The program was succeeded in 2010 by the Choice Neighborhoods program, discussed below.

The primary objective of HOPE VI was to improve the living conditions of residents in "severely distressed public housing" and revitalize the surrounding neighborhoods by breaking up the concentration of poverty that came to define public housing. To accomplish these goals, HOPE VI grants funded the demolition of large, older public housing projects and their replacement with redesigned, mixed-income units. Given the objectives of HOPE VI, projects are smaller and lower-density than traditional high-rise public housing. They are encouraged to follow the design principles of the New Urbanism, which emphasizes a traditional urban form defined by walkable neighborhoods that contain a mix of uses and housing types (Figure 7.2).

Figure 7.2 Orchard Gardens, HOPE VI housing, Boston
Photograph by Peter Vanderwarker, Architect DHK Architects, Inc.

The other key component of HOPE VI is the provision of vouchers to allow former public housing residents to move to lower-poverty areas along with support services to promote resident "self-sufficiency" (Popkin et al. 2004, Turner, Popkin, and Rawlings 2009).

The primary facet of HOPE VI—rebuilding public housing and addressing the concentration of poverty in urban neighborhoods—follows a long line of place-based strategies that respond to poor physical conditions. However, the program also recognizes that the problems of public housing are "not simply a matter of deteriorating physical conditions, it is more importantly one of a deteriorating severely distressed population in need of services and immediate attention" (National Commission on Severely Distressed Public Housing, quoted in Popkin et al. 2004: 7). The vast majority of public housing residents targeted under HOPE VI were very low-income people of color, particularly African American women and children. Further, 88% of those who lived in surrounding neighborhoods were minorities. HOPE VI therefore intended to break up this race- and class-based segregation and spur the development of mixed-income communities (Popkin et al. 2004).

The Quality Housing and Work Responsibility Act of 1998 enabled this goal by setting limits on the proportion of very low-income households in a public housing project and by eliminating the one-to-one public housing replacement law, which made it difficult for many housing authorities to demolish any public housing. With HOPE VI Demolition Grants, cities removed much or, in some cases, all of their older housing projects. Chicago, for example, engaged in a $1.6 billion project to eliminate 20,000 units. Atlanta and New Orleans have demolished their entire stock of public housing.

With HOPE VI Revitalization Grants, cities rebuild mixed-income housing. The amount of subsidized units in a HOPE VI project varies from as high as 80% very low-income units to only about 20% (Oakley et al. 2011, Tach, Pendall, and Derian 2014). Additionally, market rate units may be occupied by HCV holders. As a result, these mixed-income developments may not always contain a wide income mix. Alongside this, critics charge that the demolition and mixed-income rebuilding emphasis of HOPE VI has led to a decline in affordable rental units. HOPE VI has led to the construction of approximately 108,000 new units, but only 57,000 are affordable. By 2010, the program had resulted either directly or indirectly in the demolition of approximately 150,000 public housing units across the country (Oakley et al. 2011).

The primary success of HOPE VI has been the revitalization of the neighborhoods where the public housing was located. The program has enabled many cities to demolish their aging public housing and build complexes that considerably improved conditions in the immediate area. Projects not only include new housing, but also may provide pedestrian amenities, landscaping, parks and open space, and new street grids that improve integration of the housing complexes within the surrounding neighborhood.

HOPE VI projects, particularly those that adhered to the mixed-income requirements or that are in strong housing markets, have also sparked new private sector housing and retail development in the surrounding area. HOPE VI neighborhoods also exhibit higher incomes per capita, lower vacancy rates, and higher property values than prior to investment (Castells 2010, Voith and Zielenbach 2010). A criticism of this success, however, has been that in promoting mixed-income communities in high-poverty areas, HOPE VI is ultimately encouraging their gentrification and the displacement of the poor through the "vouchering out" of public housing residents (Bridge, Butler, and Lees 2012, Goetz 2013). Atlanta's HOPE VI projects were linked to center city redevelopment for the Olympic Games, Chicago's Cabrini Green was adjacent to some of the city's highest property values, and many other cities including Houston, New Orleans, and Washington, DC have seen property values skyrocket around former public housing sites (Goetz 2013, Keating 2000).

While HOPE VI may contribute to better neighborhood conditions, as Goetz and Chapple (2010) document in their comprehensive review of dispersal policy impacts, the program often does not produce significant improvements for former public housing residents. First, residents often do not move to better neighborhoods, but frequently remain near former public housing sites in the central city. This occurs for a number of reasons including the lack of affordable housing outside city centers, landlord discrimination, and voucher restrictions. They may also continue to live in poorly served neighborhoods because of place-based attachments and social ties.

Former public housing residents used vouchers to move out of some of the worst living conditions in the country and into neighborhoods with considerably better conditions. New neighborhoods possess significantly lower rates of poverty, unemployment, and public assistance, and residents report improvements in their housing and fewer neighborhood problems. Still, this does not necessarily translate into movement to low-poverty areas—most HOPE VI relocatees moved to neighborhoods with over 30% poverty and those that were only slightly more racially diverse than their previous residence. In other words, HOPE VI vouchers helped former public housing residents to move to better areas because public housing conditions were so poor and not necessarily because their new neighborhoods have the services and amenities residents need.

While residents often experience neighborhood improvements, individual improvements seem to be negligible. Studies show that HOPE VI relocatees had no improvement in terms of either employment or income levels. Similarly, studies do not find improved levels of physical health or education and some even report worsened conditions. Contrary to the social mixing literature, residents report few signs of social integration in their new neighborhoods and few create new social ties and supportive relationships, although youth are more likely to build social connections than adults (Goetz and Chapple 2010).

It is difficult to pinpoint precisely the explanation for the weak record of HOPE VI-related vouchers. While it may be due to the theory of social mixing itself, it could instead be a product of poor implementation and policy design. It is possible, too, that we simply need a longer evaluation period. Many of the studies that report on HOPE VI outcomes were conducted ten years or less following residential moves. However, although residents may move to better places, a variety of factors seem to keep them in high-poverty areas. The history of urban development and economic change form structural inequalities that are difficult, though not impossible, to overcome (Boxes 7.1 and 7.2).

BOX 7.1 MIXED-INCOME HOUSING AND GENTRIFICATION: DO LOW-INCOME HOUSEHOLDS BENEFIT?

Many of the neighborhoods where HOPE VI and other mixed-income projects are built have experienced a variety of neighborhood improvements. Residents gain high-quality housing and better access to retail, groceries, public transit, and new public amenities such as street trees and parks. Property values often increase and crime declines. At the same time, however, because the mixed-income approach seeks to introduce higher-income households into a neighborhood, it is by design a vehicle for gentrification. As a result, many projects have been associated with the displacement of poor households from their neighborhoods. Is it possible to support mixed-income communities that bring benefits to low-income households?

Mixed-income housing policy must do a better job of addressing residential displacement in two ways. First, policy must address direct displacement due to the construction of a mixed-income complex. Mixed-income housing construction creates fewer low-income units than previously existed in the projects that they replace. This situation can be ameliorated by enforcing a one-to-one replacement rule on the demolition of existing affordable units at or nearby the site.

Local and federal government can also pursue phased redevelopment. This would allow more residents to remain in the neighborhood during construction rather than relocating elsewhere with a housing voucher. To date, only two individual HOPE VI projects, one in

Chicago and one in Pittsburgh, have phased in development, and in Chicago this was due to a court order (Goetz 2013). The success of phased development has been demonstrated on a small scale by BC Workshop, which developed the "holding house" concept for block-level revitalization of a distressed neighborhood in Dallas, TX (HUD 2013b). The holding house provides a temporary home for residents while each house on the block is individually renovated. Using the holding house, the community design group redeveloped a single block over two years while allowing all residents to continue living there.

Mixed-income housing policy must also address the separate problem of displacement due to the subsequent gentrification of the neighborhood. HOPE VI and similar mixed-income projects are anchors for community revitalization and therefore are integral to ensuring that neighborhood change benefits all. Local, state, and federal governments can develop and implement strategies to manage the negative effects of a new project. For one, they can approach mixed-income housing on a neighborhood rather than a project-based scale. In areas where signs of gentrification are present prior to project construction, projects can incorporate higher ratios of affordable units. In other cases, such as neighborhoods with highly polarized incomes, the focus can turn toward middle-income units.

Another approach to mitigate displacement is to set aside affordable units for future purchase by public housing and neighborhood residents so that those who have the opportunity and interest in buying a home do not have to leave the neighborhood as rents increase (Chaskin, Khare, and Joseph 2012). One way to achieve this goal in gentrifying markets is to establish a community land trust (CLT) or limited-equity coop. CLTs allow occupants to own their home while a nonprofit organization retains title to the land. This strategy allows the landowner to keep the property affordable while allowing homeowners to build equity on the value of their home. Similarly, in a limited-equity coop, residents own shares of the housing complex and can sell their shares at set prices that balance affordability with equity growth.

Affordable housing retention strategies must exist alongside policies that address the broader challenges of gentrification and poverty in mixed-income communities. Fraser, Chaskin, and Bazuin (2013) suggest that employment, social services, and neighborhood programs are crucial. For example, employment programs such as the Transitional Jobs program in Chicago assist residents in improving their job prospects and attain a living-wage job by providing interview training, counselling, and three months of subsidized employment for those with little work experience. Alongside employment, social support services such as childcare and afterschool programs are important to help single mothers and working parents maintain employment. Finally, because differences in culture and class may be pronounced in gentrifying neighborhoods, programs to encourage community engagement and maintain place attachment are important. In this regard cultural festivals, community gardening, and historic preservation can all be important components of neighborhood revitalization in mixed-income communities.

BOX 7.2 NEW YORK CITY'S HOUSING PLAN: INCENTIVE ZONING AND THE AFFORDABLE HOUSING CONFLICT

Many cities face an affordable housing conflict: they need to increase their stock of affordable housing to meet the needs of lower-income groups, but are dependent on developers who focus on higher-end housing to enhance their return on investment. To tackle this "crisis of affordability," New York City Mayor Bill de Blasio recently unveiled a new ten-year housing plan (City of New York 2014). Calling it the largest city-level housing program in US history, the mayor seeks to build and preserve 200,000 affordable units for low- and middle-income households over the next ten years.

The New York plan sets out an array of initiatives including programs to preserve rent-regulated units, protect tenants, update zoning codes, and create more economically diverse neighborhoods. The plan creates two new programs, the Neighborhood Construction Program (NCP) and the New Infill Homeownership Opportunities Program (NIHOP), to encourage small developers to construct affordable housing and increase homeownership on vacant properties.

At the center of the plan is the Inclusionary Housing Program, which will require developers to include permanent affordable units for low- or moderate-income households in new residential projects. In exchange for dedicating a percentage of units as affordable, usually 20%, the developer will receive a density bonus (the right to build more housing units in larger buildings than allowed in a designated zone). This incentive zoning (IZ) program is a fairly common tool applied by cities to increase their stock of affordable housing and create economically diverse communities. IZ programs may also include expedited permit processes, fee waivers, and reduced zoning standards and parking requirements (Schuetz, Meltzer, and Been 2009). Some IZ policies allow construction of affordable units off site or cash in lieu of construction. As of 2009, an estimated 80,000–120,000 affordable units have been built under 300–500 local inclusionary zoning ordinances. One of the most successful programs, in Montgomery County, MD, has produced approximately 13,000 units since its inception (Calavita and Mallach 2009).

In most cases, including New York's current program, IZ programs are voluntary. Some challenge that voluntary programs do not have a major impact on the construction of affordable housing. Both New York's program and Seattle's IZ, for instance, do not keep pace with affordable housing needs, generating less than 2% of all multifamily units developed over roughly the past decade (Downtown Seattle Association 2014, Navarro 2013). Further, housing advocates criticize some IZ programs such as those in San Francisco and San Jose, CA for allowing households that earn up to 120% of the area median income (AMI) to qualify for affordable units rather than targeting low- and very low-income families. The New York plan targets a mix of incomes from extremely low- to middle-income households with the bulk targeted at low-income families (58%) and just 11% for those earning 120% AMI.

New York wants to follow a handful of other cities, including Boston, Denver, San Francisco, Washington, DC and San Jose, CA, that require developers to provide affordable housing in new residential projects. The success of any IZ program is dependent on generating new development. When this occurs, IZ programs can create more affordable housing in new development, which typically have good access to services and amenities. However, detractors claim that mandatory IZs discourage development because they reduce developer profits and, as a result, will actually slow the construction of new affordable housing. Can New York achieve its ambitious housing goals and encourage developers to build more affordable housing in economically diverse communities?

OTHER ROUTES TOWARD NEIGHBORHOOD REVITALIZATION: CHOICE NEIGHBORHOODS AND EMPOWERMENT ZONES

Many believe that housing is not enough to break up high-poverty areas and spur neighborhood revitalization. Some argue that we need more market-oriented strategies focused on attracting businesses and providing employment opportunities in low-income neighborhoods. Others assert that the focus should be on comprehensive community initiatives (CCI), which take on a more wide-ranging neighborhood revitalization agenda. As the Obama administration's Neighborhood Revitalization Initiative (White House 2012) states, "the interconnected challenges in high-poverty neighborhoods require interconnected solutions. Struggling schools, little access to capital, high unemployment, poor housing, persistent crime and other challenges feed into and perpetuate each other, intensifying challenges for residents." CCIs attempt to address these challenges through place-based efforts across multiple sectors in coordination with residents and community organizations.

A COMPREHENSIVE COMMUNITY INITIATIVE: CHOICE NEIGHBORHOODS

The Choice Neighborhoods program is currently the most high-profile CCI in the US. The program expands HOPE VI by integrating housing into comprehensive place-based revitalization projects. The program is part of the larger White House Neighborhood Revitalization initiative.[2] The Neighborhood Revitalization initiative is geared toward improving engagement across federal agencies such as the Department of Economic Development, Department of Justice, and HUD, as well as fostering partnerships among local community organizations, governmental agencies, property developers, and business owners in creating "neighborhoods of opportunity" (Smith 2011, Urban Institute 2013, White House 2012).

Choice Neighborhoods is focused around three core goals of housing, people, and neighborhoods. As with HOPE VI, projects must focus on replacing distressed public housing with mixed-income units. Additionally, they must design education and support services geared toward improving "intergenerational mobility." Third, programs tackle neighborhood-level issues by encouraging public–private partnerships and investment in distressed areas that focus on coordinating programs around employment and business development, public safety, transportation, and the development of amenities (HUD n.d.-a).

Choice Neighborhoods is a competitive program that awards two types of grants. Planning grants range from $167,000 to $500,000 to support the development of neighborhood revitalization plans ("Transformation Plans") that address the three core program goals. Implementation grants are considerably larger, up to $30 million, and fund implementation of the transformation plans. Since 2010, HUD has awarded 56 planning grants totaling $16.9 million and 13 implementation grants totaling $351 million. The award amounts, however, are actually only a small portion of the overall project budget and are intended to encourage local partnerships and move existing revitalization efforts forward. Each Choice Neighborhood site has also leveraged significant funding from multiple public and private sources.

While the main focus continues to revolve around mixed-income housing, the program frames this in the context of more comprehensive community-building strategies. The comprehensive community initiatives (CCI) movement emerged in the 1990s based on the idea that neighborhood issues cannot be effectively addressed in isolation and that community development must take a more integrated approach to economic, educational, social, physical, and cultural development at the neighborhood level. Additionally, CCIs emphasize the importance of building the capacity of community residents and local organizations to solve neighborhood issues themselves.

Evaluations of CCIs show that they deliver mixed results. Programs generally report improvements in education and family services as well as increased property values and new retail development. There is also qualitative evidence of community capacity building in CCI programs. Overall economic development has been a significant challenge, however, given that so many factors that influence employment opportunities and business development emanate from beyond the neighborhood. Another challenge is the ability to effectively break down the complex silos across services and funding streams (Kubisch et al. 2010).

It is still early to adequately evaluate even the first five Choice Neighborhood implementation sites in Boston, Chicago, New Orleans, San Francisco, and Seattle, and each project supports a broad set of activities and partnerships in diverse contexts (Urban Institute 2013). For example, the Boston award supports a partnership between City, economic and community development organizations, and public schools to revitalize a high-poverty area around HUD-assisted housing sites and focuses predominantly on the development of new housing and education programs. Projects in New Orleans and San Francisco have more diverse partners. Partners in these cities include local government, property developers, school districts, and social service agencies. The participants engage in the redevelopment of public housing sites in high-poverty areas and are creating programs for adult and child education, job training, crime prevention, health programs, and other social services. The programs also invest in new street infrastructure, retail development, and the arts. The varying scope and comprehensiveness makes comparative evaluation of Choice Neighborhood projects next to impossible but observers are closely monitoring individual projects.

EMPOWERMENT ZONES/ENTERPRISE COMMUNITIES: A MARKET-BASED APPROACH TO BRING JOBS TO LOW-INCOME NEIGHBORHOODS

Introduced in 1994 under President Clinton, Empowerment Zones/Enterprise Communities (EZ/EC) is a largely market-based approach to stimulating neighborhood revitalization. As we discussed in chapter 4, this federal urban revitalization program was designed to tackle unemployment and business development in distressed neighborhoods through competitive, incentive-based policies along with community-based participation and partnerships. The program differs from earlier state EZ programs in that it seeks to build community engagement by incorporating members of local government, businesses, residents, and community organizations in strategic planning within the zone. EZ/EC also differs from state programs in that Empowerment Zones typically concentrate on neighborhoods surrounding the downtown core, while most state EZs cover a large portion of or even the entire central city.

The EZ/EC program initiated three rounds of competitive awards (in 1994, 1998, and 2001). In the first round, 71 neighborhoods received EZ/EC awards. However, the majority of funding went to high-poverty areas in just six Empowerment Zones in Atlanta, Baltimore, Chicago, Detroit, New York, and Philadelphia. There are currently 30 active EZs although their federal tax incentives expired at the end of 2013 (US HUD). Given this distribution, many eligible neighborhoods in need of support did not receive funding through the program.

While specific activities vary, each Round I and II EZ received a ten-year, $100 million Social Services Block Grant that could be applied broadly.[3] Alongside this, businesses are eligible for a variety of federal tax credits including the EZ Employment Credit and Work Opportunity Credit, both of which allow annual tax credits for employing individuals who live and work in the zone. Businesses are also eligible for special tax deductions on property and equipment, and bond funding for property purchase and improvement (US HUD).

Evaluations of both state and federal programs show mixed results in terms of business and jobs development as well as community participation. Oakley and Tsao (2006) provide a good review of these assessment studies and report that some EZs lead to higher wages, property appreciation and lower levels of unemployment and poverty. However, other EZs exhibit no appreciable gains. For example, in Oakley and Tsao's (2006) study the Chicago and Detroit EZs maintained high poverty rates after zone designation. Moreover, studies find that job growth is just as likely to occur for reasons other than the EZ program (HUD 2001, Government Accountability Office 2006). In terms of community participation, studies find high levels during the strategic planning process, but a drop-off over the life of the program (McCarthy 1998). However, those EZs that maintained high levels of government and community involvement seem to be more successful (Rich and Stoker 2007). In short, evaluation of EZ programs seems to indicate that tax incentives alone are not sufficient to revitalize struggling communities.

A possible successor to the EZ/EC program is the recently established Promise Zone Initiative under President Obama (White House 2014). In conjunction with tax reductions modeled on the EZ/EC program, Promise Zones support partnerships between local communities and businesses primarily in job training and education and other social programs.

SUMMARY

Housing has long been central to urban revitalization policy. Since the 1990s, housing has become a primary means to facilitate income mix in high-poverty neighborhoods through two means. One approach, exemplified by Housing Choice Vouchers, focuses on dispersing residents to areas that may provide them with better living conditions and economic opportunity. The other approach, embodied by HOPE VI, is the development of mixed-income housing. Additionally, CCIs such as Choice Neighborhoods continue to present a complex approach to neighborhood revitalization and poverty deconcentration. In addition, Empowerment Zones have sought to bring more jobs to residents in job-poor, distressed neighborhoods. While concentrated poverty is a prime concern of urban revitalization policy, there continues to be a vigorous debate over whether or not spatial changes improve people's lives. Complex and complicated structural issues present a serious challenge to place-based programs. The evidence shows that deconcentration policies are better at revitalizing—or gentrifying—places than at helping to lift people out of poverty.

STUDY QUESTIONS

1 Why is housing a key focus of urban revitalization policy? Do you feel other approaches might be more useful?
2 Should the federal government return to funding affordable public housing projects or continue to fund mixed-income housing?
3 Do you think that Choice Neighborhoods will result in significant place- and people-based improvements over other urban revitalization programs? Consider precedents from other programs to support your position.
4 Identify the location of houses and apartments that accept Housing Choice Vouchers, a Low Income Housing Tax Credit development, and a HOPE VI project (if there is one in your city or town). Where are they located? Describe the neighborhood conditions, the services available (such as grocery stores), and the location in relationship to job centers. Do the residents have high-quality neighborhoods and access to jobs, goods, and services? Why, or why not?

NOTES

1 The latter includes areas where housing costs are high relative to incomes and the former includes tracts in which 50% of households have incomes below 60% of the area median or a poverty rate of 25% or more.
2 Along with Choice Neighborhoods, the Neighborhood Revitalization Initiative includes the Promise Neighborhoods, Byrne Criminal Justice Innovation, and Community Health Center programs (White House 2012).
3 EC awardees received a $3 million block grant and no tax credits. Round III EZs and ECs did not receive the block grants.

REFERENCES

Basolo, Victoria. 2013. "Examining mobility outcomes in the Housing Choice Voucher Program: neighborhood poverty, employment, and public school quality." *Cityscape* 15(2): 135–153.

Bridge, Gary, Tim Butler, and Loretta Lees. 2012. *Mixed communities: gentrification by stealth?* Bristol, UK: Policy Press.

Briggs, Xavier de Souza, Susan J Popkin, and John Goering. 2010. *Moving to opportunity: the story of an American experiment to fight ghetto poverty*. New York: Oxford University Press.

Calavita, Nico and Alan Mallach. 2009. "Inclusionary housing, incentives, and land value recapture." *Land Lines* 21(1): 15–21.

Castells, Nina. 2010. "HOPE VI neighborhood spillover effects in Baltimore." *Cityscape* 12(1): 65–98.

Chaskin, Robert J and Mark L Joseph. 2011. "Social interaction in mixed-income developments: relational expectations and emerging reality." *Journal of Urban Affairs* 33(2): 209–237. doi: 10.1111/j.1467-9906.2010.00537.x

Chaskin, Robert, Amy Khare, and Mark Joseph. 2012. "Participation, deliberation, and decision making: the dynamics of inclusion and exclusion in mixed-income developments." *Urban Affairs Review* 48(6): 863–906.

City of New York. 2014. "Housing New York: a five borough, ten year plan." New York: City of New York.

Dawkins, Casey. 2013. "The spatial pattern of low income housing tax credit properties: implications for fair housing and poverty deconcentration policies." *Journal of the American Planning Association* 79(3): 222–234. doi: 10.1080/01944363.2014.895635

Dewan, Shaila. 2014. "In many cities, rent is rising out of reach of middle class." *New York Times* April 14. www.nytimes.com/2014/04/15/business/more-renters-find-30-affordability-ratio-unattainable.html?_r=0

Downtown Seattle Association. 2014. "Seattle's Incentive Zoning Program: facts and conclusions." www.downtownseattle.com/assets/2014/02/City-Housing-Forum-IZ-Fact-Sheet.pdf

Fraser, James Curtis, Ashley Brown Burns, Joshua Theodore Bazuin, and Deirdre Áine Oakley. 2012. "HOPE VI, colonization, and the production of difference." *Urban Affairs Review* 49(4): 525–556.

Fraser, James C, Robert J Chaskin, and Joshua Theodore Bazuin. 2013. "Making mixed-income neighborhoods work for low-income households." *Cityscape* 15(2): 83–100.

Goetz, Edward G. 2003. "Housing dispersal programs." *Journal of Planning Literature* 18(1): 3–16. doi: 10.1177/0885412203251339

Goetz, Edward G. 2013. *New Deal ruins: race, economic justice, and public housing policy*. New York: Cornell University Press.

Goetz, Edward G and Karen Chapple. 2010 "You gotta move: advancing the debate on the record of dispersal." *Housing Policy Debate*, 20(2): 209–236. doi: 10.1080/10511481003779876

Government Accountability Office. 2006. *Empowerment Zone and Enterprise Community Program: improvements occurred in communities, but the effect of the program is unclear*. GAO-06-727. Washington, DC: Government Accountability Office.

Haas Institute. 2015. "Haas Institute co authors amicus brief submitted to Supreme Court on fair housing case in Texas." Berkeley, CA: Haas Institute. http://diversity.berkeley.edu/publication-SCOTUS-amicus-brief-scholars-texas-housing-case

Hackworth, Jason. 2007. *The neoliberal city: governance, ideology, and development in American urbanism*. New York: Cornell University Press.

HUD. 2001. *Interim assessment of the Empowerment Zones and Enterprise Communities (EZ/EC) program: A progress report*. Washington, DC: US Department of Housing and Urban Development.

HUD. 2013a. *Worst case housing needs 2011: Report to Congress*. Washington, DC: US Department of Housing and Urban Development.

HUD. 2013b. "Dallas, Texas: Congo Street green initiative provides important lessons in community revitalization." Washington, DC: US Department of Housing and Urban Development. www.huduser.org/portal/casestudies/study_04152013_1.html

HUD. n.d.-a. "Choice Neighborhoods." Washington, DC: US Department of Housing and Urban Development. http://portal.hud.gov/hudportal/HUD?src=/program_offices/public_indian_housing/programs/ph/cn

HUD. n.d.-b. "List of current Empowerment Zones and updated contact information." Washington, DC: US Department of Housing and Urban Development. http://portal.hud.gov/hudportal/HUD?src=/program_offices/comm_planning/economicdevelopment/programs/rc/ezcontacts

HUD. n.d.-c. "Low-Income Housing Tax Credits." Washington, DC: US Department of Housing and Urban Development. www.huduser.org/portal/datasets/lihtc.html

Jargowsky, Paul. 2003. "Stunning progress, hidden problems: the dramatic decline of concentrated poverty in the 1990s." Washington, DC: Brookings Institution. www.brookings.edu/research/reports/2003/05/demographics-jargowsky

Joint Center for Housing Studies. 2013. *The state of the nation's housing*. Cambridge, MA: Joint Center for Housing Studies of Harvard University. www.jchs.harvard.edu/research/state_nations_housing

Jones, Katherine T and Jeff Popke. 2010. "Re-envisioning the city: Lefebvre, Hope VI, and the neoliberalization of urban space." *Urban Geography* 31(1): 114–133. doi: 10.2747/0272-3638.31.1.114

Joseph, Mark L. 2006. "Is mixed-income development an antidote to urban poverty?" *Housing Policy Debate* 17(2): 209–234. doi: 10.1080/10511482.2006.9521567

Keating, Larry. 2000. "Redeveloping public housing." *Journal of the American Planning Association* 66(4): 384–397. doi: 10.1080/01944360008976122

Kneebone, Elizabeth and Alan Berube. 2013. *Confronting suburban poverty in America*. Washington, DC: Brookings Institution.

Kubisch, Anne C, Patricia Auspos, Prudence Brown, and Thomas Dewar. 2010. "Community change initiatives from 1990–2010: accomplishments and implications for future work." *Community Investments* 22(1): 8–36.

Massey, Douglas S and Nancy A Denton. 1993. *American apartheid: segregation and the making of the underclass*. Cambridge, MA: Harvard University Press.

McCarthy, John. 1998. "US urban empowerment zones." *Land Use Policy* 15(4): 319–330.

McClure, Kirk. 2010. "The prospects for guiding housing choice voucher households to high-opportunity neighborhoods." *Cityscape* 12(3): 101–122.

Navarro, Mireya. 2013. "Report finds a city incentive is not producing enough affordable housing." *New York Times* August 15. www.nytimes.com/2013/08/16/nyregion/report-finds-a-city-incentive-is-not-producing-enough-affordable-housing.html?_r=0

Oakley, Deirdre and Hui-Shien Tsao. 2006. "A new way of revitalising distressed urban communities? Assessing the impact of the federal Empowerment Zone program." *Journal of Urban Affairs* 28(5): 443–471. doi: 10.1111/j.1467-9906.2006.00309.x

Oakley, Deirdre, Chandra Ward, Lesley Reid, and Erin Ruel. 2011. "The poverty deconcentration imperative and public housing transformation." *Sociology Compass* 5(9): 824–833. doi: 10.1111/j.1751-9020.2011.00405.x

Polikoff, Alexander and Clarence Page. 2007. *Waiting for Gautreaux: a story of segregation, housing, and the black ghetto*. Evanston, IL: Northwestern University Press.

Popkin, Susan J, Bruce Katz, Mary K Cunningham, Karen D Brown, Jeremy Gustafson, and Margery Austin Turner. 2004. "A decade of HOPE VI: research findings and policy challenges." Washington, DC: Urban Institute. www.urban.org/url.cfm?ID=411002&renderforprint=1&CFID=78071151&CFTOKEN=47422001&jsessionid=f03070a4cb957b9301cd80707de14266c247

Porter, Michael E. 1997. "New strategies for inner-city economic development." *Economic Development Quarterly* 11(1): 11–27.

Rich, Michael J and Robert P Stoker. 2007. "Governance and urban revitalization: lessons from the urban empowerment zones initiative." Paper presented at Conference on "A Global Look at Urban and Regional Governance: The State–Market–Civic Nexus." Atlanta, GA: Emory University.

Rosen, Kenneth T and Ted Dienstfrey. 1998. "The economics of housing services in low income neighborhoods." In *Program on Housing and Urban Policy*. Working Paper Series. Berkeley, CA: Institute of Business and Economic Research and Fisher Center for Real Estate and Urban Economics.

Schuetz, Jenny, Rachel Meltzer, and Vicki Been. 2009. "31 flavors of inclusionary zoning: comparing policies from San Francisco, Washington, DC, and suburban Boston." *Journal of the American Planning Association* 75(4): 441–456.

Schwartz, Alex. 2011. "Low Income Housing Tax Credits." In *Fair and affordable housing in the U.S.*, edited by Robert Mark Silverman and Kelly L Patterson. Leiden, The Netherlands: Brill.

Smith, Robin E. 2011. *How to evaluate Choice and Promise Neighborhoods*. Washington, DC: Urban Institute. www.urban.org/publications/412317.html

Tach, Laura, Rolf Pendall, and Alexandra Derian. 2014. *Income mixing across scales: rationale, trends, policies, practice, and research for more inclusive neighborhoods and metropolitan areas*. Washington, DC: Urban Institute.

Turner, Margery Austin, Susan J Popkin, and Lynette Rawlings. 2009. *Public housing and the legacy of segregation*. Washington, DC: Urban Insitute.

Urban Institute. 2013. *Developing Choice Neighborhoods: an early look at implementation in five sites – Interim report*. Washington, DC: US Department of Housing and Urban Development Office of Policy Development and Research.

Voith, Richard and Sean Zielenbach. 2010. "Hope VI and neighborhood economic development: the importance of local market dynamics." *Cityscape* 12(1): 99–132.

White House. 2012. "Neighborhood Revitalization Initiative." www.whitehouse.gov/administration/eop/oua/initiatives/neighborhood-revitalization

White House. 2014. "Fact Sheet: President Obama's Promise Zones Initiative." www.whitehouse.gov/the-press-office/2014/01/08/fact-sheet-president-obama-s-promise-zones-initiative

Wilson, William Julius. 1987. *The truly disadvantaged: the inner city, the underclass, and public policy*. Chicago, IL: University of Chicago Press.

Wilson, William Julius. 1996. *When work disappears: the world of the new urban poor.* 1st Ed. New York: Vintage Books.

8
Reconfiguring the suburbs
Strategies for commercial revitalization and rebuilding sprawl

For much of the twentieth century, the suburbs served as the symbolic antithesis of urban decline. The city was stigmatized as a place of blight, poverty, and fear, while homeownership in the suburbs represented the American dream (Schafran 2013). Indeed, for many, suburban homeownership has provided a route to opportunity and upward social mobility. Property ownership has been the primary way that middle-income households acquired and transferred wealth from generation to generation. The suburbs offered the chance for many to become homeowners and move to a neighborhood with good schools and a safe, tight-knit, and family-friendly community. Nevertheless, the suburbs have their critics, who deride them as dull, homogenous, and environmentally harmful (Duany, Plater-Zyberk, and Speck 2010, Kunstler 1994). Despite this, the outer suburbs continue to expand, suggesting that the opportunities still appeal to many people.

As the exurbs expand and money increasingly goes toward redevelopment in the urban core, inner-ring suburbs built in the early and mid-twentieth century are facing serious challenges (Hanlon 2009, Hanlon, Vicino, and Short 2006, Kneebone and Berube 2013, Puentes and Orfield 2002, Puentes and Warren 2006, Short, Hanlon, and Vicino 2007). Scholars and public media alike increasingly have found that "decline is a new suburban reality":

> After two centuries of development, US suburbia has evolved into a new reality that includes continued growth and prosperity *and* decline and poverty. Gated communities, McMansions, and the supersized subdivision exist alongside much poorer suburbs struggling with issues of blight, fiscal stress, income decline, increasing poverty and housing deterioration.
>
> (Hanlon 2009: 12)

The challenges facing suburban revitalization are familiar but distinct. Suburban revitalization requires paying attention to its auto-oriented urban form, the predominance of residential development, and underused shopping centers and big box stores. These are distinct features when compared with revitalization efforts in the urban core.

These conditions have resulted in a call for "retrofitting suburbia." Dunham-Jones and Williamson (2008) and Tachieva (2010) among others argue that by *redesigning* the suburbs around higher densities, improving transit, and addressing other issues related to suburban form, we can create more environmentally and socially sustainable suburbs.

Another important factor for suburban revitalization involves recognizing that many suburban areas have experienced major demographic changes. Their diversity has increased across many characteristics including income, race, ethnicity, languages spoken, and country of birth. In fact, the suburbs are now more diverse than the US as a whole (Frey, Berube, Singer, and Wilson 2009, Orfield and Luce 2013). The growing diversity is a potential asset for suburban revitalization. Many so-called declining suburbs have in fact been transformed into thriving business districts at the hands of independent immigrant communities and entrepreneurs. In such cases, rather than viewing the aging, obsolete buildings and inner-ring suburban commercial corridors as declining and unsustainable, they offer affordable, adaptable spaces. Unfortunately, however, while diversity brings such opportunities, many suburban neighborhoods still remain socially segregated and isolated from economic opportunity.

This chapter dives into these contrasting views and approaches to suburban revitalization. We begin by discussing how suburbs differ from other urban areas and explore the challenges facing inner-ring suburbs and suburban revitalization. We then discuss the leading trends and strategies shaping approaches to retrofitting suburbs including programs to transform underutilized shopping centers and create vibrant, diverse, and dense communities in formerly low-density landscapes.

SUBURBAN MYTHS AND REALITIES

Since the nineteenth century, suburbs have stood as the antithesis of the city. Suburbs were the safe realm of domestic life compared with the dangers associated with the fast-paced, business-oriented central city. Throughout the twentieth century, suburbs were regularly portrayed as clean and safe. They became places with good schools and easy access (by car) to a range of places, while downtowns and central city neighborhoods were labeled as dangerous and crowded with poor schools and traffic congestion.

Suburban realities are considerably more complex. No single perspective accurately describes the suburbs because they encompass so many different places and mean many different things to their diverse residents. While most people understand a suburb to be a residential area located outside of the urban core, the consensus stops there. Is it an American dream of homeownership or a stifling homogeneity? Is sprawl an environmental disaster or is it green, large-lot neighborhoods that are the best place to raise kids? Are suburban residents white, wealthy, and exclusive or diverse and open to difference?

In fact, there is no agreed-upon definition of the suburb, let alone the inner-ring suburbs that are the focus of this chapter. Nevertheless, we can identify some basic features that help explain their situation. Suburbs refer to predominantly residential neighborhoods located within metropolitan areas but outside the central city. They might be incorporated cities or unincorporated parts of a county. For example, unincorporated Metairie, LA, a suburb of New Orleans located in Jefferson Parish, has a 2010 population of over 130,000. Kenner, LA, another New Orleans suburb in Jefferson Parish, is an incorporated city with over 66,000 residents. Both would be considered suburbs of New Orleans.

Suburbs are defined by both their residential character and their physical dimensions. In particular, suburbs are distinguished by low-density patterns of development and single-use zoning comprised primarily of single-family homes on large blocks and poor street connectivity (Forsyth 2012). Although many commercial and industrial districts have been developed outside of the urban core, they are separate from most suburban residential neighborhoods, and mixed-use areas are rare. Until relatively recently, suburban areas lacked major concentrations of employment.

These general development patterns produce urban sprawl. According to *Measuring Sprawl 2014* (Ewing and Hamidi 2014), the extent to which different places exhibit sprawling development patterns can be evaluated on four dimensions—residential and employment density, land-use mix, street connectivity, and activity centering. Based on these criteria, most suburban areas qualify as sprawling. However, there are different types of sprawl. Some areas exhibit uniformly low densities. Others are characterized by a pattern of leapfrog development, a process by which developers bypass built-up inner-suburban locations for less expensive and easier-to-develop greenfield property. "Edge city" areas may contain relatively dense employment concentrations surrounded by low-density residential development. Still other areas are characterized by uniformly distributed density or "dense sprawl."

While many people opt to live in suburbs, sprawling development comes with numerous costs. For decades, critics have charged that the typical characteristics of suburbia—low density, single-use zoning, auto dependence, and often uncoordinated development—have produced a host of environmental problems (Duany et al. 2010). These include poor air and water quality, the loss of open space and agricultural land, and harm to natural habitats.

Because of these characteristics, and because mass transit is less frequent or convenient, residents in sprawling suburbs become auto dependent. This tends to require residents to make longer and more frequent driving trips than if their needs were clustered in one place. This too has an environmental impact. More vehicle miles traveled (VMT) means unnecessary energy consumption and greenhouse gas (GHG) emissions that contribute to global warming.

Another major critique concerns suburban leapfrog development. The construction of new residential development, malls, and lifestyle centers on the suburban periphery draws businesses and jobs from small downtowns and older retail strips, creating a surplus of failing greyfield properties or "dead malls" while often resulting in an oversupply of retail elsewhere. Leapfrog development and large-format retail vacancy cause not only environmental problems in that they contribute to longer drive-times through sprawling suburbs, but also economic problems because the vacancies often damage the tax and employment base of inner-suburban municipalities. Further, the increasing amount of large, vacant parcels and visual blight both discourages new infill development and can set off a chain reaction of store closures from smaller retailers that depend on larger anchor stores for traffic. This, in turn, can suppress municipal property values and force residents of many communities to drive further from home to meet their basic needs.

Sprawl has also been associated with health issues. Lower-density, less connected places tend to be associated with higher rates of obesity, heart disease, and diabetes than those that are more compact and better connected (Ewing and Hamidi 2014, Marshall, Piatkowski, and Garrick 2014).

Finally, suburban development has long been criticized for its social costs including a lack of community connections (Jacobs 1961). More recently, however, as people have begun to appreciate suburban diversity, the focus has turned toward recognizing that postwar suburban development patterns do not support changing demographics and new consumer demands. Suburban neighborhoods today house fewer traditional households comprised of two parents and children and are home to growing numbers of singles, unmarried couples, and aging baby-boomers (Frey et al. 2009).

THE STATE OF INNER-RING SUBURBS

Scholars increasingly recognize the diversity of suburban places and have begun to eschew the central city/suburb dichotomy in favor of differentiating between different types of suburbs based on geography and age of housing stock (Hanlon 2009, Lee and Green Leigh 2007). Outer-ring suburbs are farthest from the urban core, contain newer housing stock, and are more prosperous. As they continue to grow, those areas that were booming in the postwar years—the inner-ring suburbs—are now gaining population at a much slower rate or even losing people (Green Leigh and Lee 2005, Puentes and Warren 2006).

Inner-ring or first suburbs are located near or adjacent to the primary urban core or downtown and contain housing stock built before 1970. There are robust inner-ring suburbs that contain older housing with historic character built prior to 1940. Many of these historic neighborhoods are closer to the urban core and have experienced gentrification as more people move to the urban core. However, according to Hanlon (2009), two-thirds of all distressed suburbs are located in the inner ring. Those most likely to be struggling with symptoms of decline contain a majority of housing stock built following World War II, between 1950 and 1969, when developers laid out very large subdivisions of single-family homes.

This pattern of disinvestment reflects the continued decentralization of population and employment to the outer-ring suburbs. Middle- and upper-income households are attracted to the urban fringe by newer and larger housing on larger lots. Some of these households also seek out residences in the urban core. As existing inner-ring suburban residents move elsewhere in search of better housing options, this opens up the supply of housing for residents attempting to move out of extremely poor central city areas and into comparatively better housing in the inner-ring suburbs. It also provides an option for those who are displaced as urban core neighborhoods gentrify (Lee and Green Leigh 2007, Orfield 1997).

A key issue that defines suburban distress is mounting suburban poverty, a growing trend in the US as well as Europe, Australia, and Canada (Randolph and Tice 2014). In the US, there has been a dramatic increase in the number of suburban households with incomes below the federal poverty line. Although the proportion of low-income residents is still higher in the city, suburban residents in poverty now exceed those living in the urban core by 1.5 million. Additionally, poverty rates are climbing faster there than in the urban core (Frey et al. 2009, Kneebone and Berube 2013). According to Kneebone and Berube (2013), suburbs in the 100 largest US metros have experienced a near 64% increase in poverty while central cities increased by less than half of this rate over the first decade of the 2000s. While trends vary by specific places, a look at the Urban Institute's Poverty Map in 1980 and 2010 illustrates the movement of poverty out of Chicago's urban core and its concentration in inner-suburban areas (Figure 8.1).

Declining suburbs also face shrinking incomes. Over the last decades of the 2000s, studies show that median incomes declined faster in many suburbs than in the surrounding central cities and that postwar inner-ring suburbs were the worst off in this regard (Green Leigh and Lee 2005, Lucy and Phillips 2000). Moreover, poor and wealthy suburbs have become more segregated by income since the 1980s, particularly in growing southwest cities such as Houston, Los Angeles, and Miami (Hanlon 2009).

Inner-suburban housing and subdivisions are aging and in need of repair, which results in stagnating or depreciating property values. The neighborhoods also wrestle with aging infrastructure and commercial corridors, lack of public transportation, and underfunded public schools (Puentes and Orfield 2002). In some places, declining commercial corridors and abandoned malls pose serious problems for localities as well. Labels such as "slumburbia" have emerged that depict suburban neighborhoods that have become dilapidated and abandoned, much like those descriptions formerly reserved for challenged central city neighborhoods (Schafran 2013: 135).

As a result of these conditions, declining suburbs are fiscally challenged. According to Puentes and Orfield (2002) over two-thirds of the suburban population are living in areas experiencing some form of fiscal stress. Drops in property tax revenues and a weak commercial tax base make it difficult to deal with social service issues and to repair and upgrade public infrastructure. Budget pressures are exacerbated by the high level of municipal fragmentation and competition that exists in many regions, where dozens if not hundreds of small cities with their own land use, zoning, and taxing powers vie for new investment and residents. They also often lack the organizational capacity in the form of local nonprofit service providers to help address their challenges. In this context, revitalization by individual suburban cities is exceedingly difficult.

Figure 8.1 Chicago poverty map, 1980 (top) and 2010 (bottom); one dot represents 20 people with income below the poverty line

Maps reprinted with permission of the Urban Institute

REVITALIZING SUBURBIA

While many inner-ring suburbs are in dire need of revitalization assistance, they often exist in a "policy blindspot" (Puentes and Orfield 2002). Many suburban municipalities are ineligible for federal community development funding such as CDBGs or Enterprise Zone funds because they are too small or, although struggling, do not meet minimum poverty thresholds. To compound the problems, residents in many suburban cities lack the social services and public transit that can help connect them to better employment opportunities. Poverty is also less visible in the sprawling suburbs compared with the dense urban core.

Many people have advocated for greater federal and state intervention (Hanlon 2009, Puentes and Orfield 2002, Vicino 2008), but few initiatives have been developed to address suburban decline. Beyond the unsuccessful attempt in 2005 to pass the Suburban Core Opportunity Restoration and Enhancement (SCORE) Act in Congress, which would have provided tax incentives and a $250 million Reinvestment Fund to spur the revitalization of aging suburbs, there have been no federal attempts to directly address suburban decline.

State-level smart growth policies and "fix-it-first" infrastructure programs are more common. These incentive-based programs attempt to alter sprawling urban growth patterns and preserve natural habitat by directing development to already built-up areas in the urban core and inner-ring suburbs. A number of states, including California, Delaware, Florida, Georgia, Maryland, New Jersey, Oregon, Pennsylvania, Rhode Island, Tennessee, Vermont, Washington, and Wisconsin maintain some type of targeted area or smart growth funding program aimed at rehabilitating existing infrastructure and encouraging infill development at least partly in the inner suburbs. One of the most ambitious programs is California's "anti-sprawl" bill (see Box 8.1).

In some regions, coalitions of inner-suburban municipalities come together to lobby state and federal entities to obtain funding for infrastructure needs and address social and economic concerns. For example, the Northeastern Ohio First Suburbs Consortium has worked with private foundations to fund studies of commercial retail districts in their member cities and has engaged architecture firms in marketing and renovating inner-suburban housing stock. The Consortium has also worked with Cuyahoga County to establish the Housing Enhancement Loan Program (HELP) to encourage banks to make low-interest loans (Puentes and Orfield 2002). While this and similar coalitions in Chicago, Detroit, Minneapolis, and elsewhere have had success, the formation of inner-suburban coalitions in other regions is frequently hampered by a lack of political will and inter-city competition.

Initiatives continue to be needed at the federal, state, and regional scales to implement new visions for transforming inner-ring suburbs. These visions differ from the consumption- and entertainment-oriented approach in the urban core. In the following sections we discuss three trends shaping the direction of suburban revitalization. The first trend focuses on efforts to retrofit the suburbs with mixed-use destinations, pedestrian-oriented infrastructure, and more compact development. A second and related trend revolves around adapting big box and suburban retail centers into new destinations. Finally, some areas are revitalizing without direct policy support at the hands of diverse residents who find opportunities in formerly declining suburbs.

RETROFITTING SUBURBS

A growing number of architects and planners have turned their attention to redesigning and "retrofitting suburbia" (Dunham-Jones and Williamson 2008, Grant et al. 2013, Tachieva 2010, Williamson 2013). The call for retrofitting the suburbs stems from critiques of postwar suburban development that we discuss above, particularly as they relate to the need for more sustainable suburban form and the failure of suburban development to meet changing demographics. This nascent movement is focused on urbanizing the suburbs primarily through design-based strategies that are strongly aligned with the new urbanism.[1] Proponents call for the development of compact, mixed-use, and pedestrian-oriented destinations, and promise "comprehensive guidance for transforming fragmented, isolated and car-dependent development into 'complete communities'" (Tachieva 2010).

Retrofitting suburbia strategies argue that to address the environmental and social issues associated with suburban sprawl, we need to redesign existing suburban development around hybrid urban nodes that are neither urban nor suburban (Dunham-Jones and Williamson 2008). The key focus is on increasing density and building mixed-use centers with compact buildings, walkable streets, and public transit access, but also suburban parking ratios. Retrofitting strategies seek to alter postwar development patterns and create denser nodes within suburban sprawl by providing multiple, smaller-scale housing options that look like traditional cities comprised of old buildings on small blocks, but are actually clustered in large-scale real estate projects or "instant cities." They seek to provide public space where diverse people may gather, but do so in the context of privatized development projects with chain retail.

The concept of remaking the suburbs around more sustainable and socially conscious design is not a new idea (see Hayden 1997). However, calls for retrofitting and repairing suburbia, addressing the environmental and social issues associated with suburban development, have taken a more central place in the public eye. The idea of remaking the suburbs has recently been promoted by popular media and development professionals as well (Talen 2011). For example, the modern home and lifestyle magazine *Dwell* recently sponsored a suburban design competition called "Reburbia," the Rausch Foundation sponsored the international Build a Better Burb design competition, and the Urban Land Institute sponsored a "Sustainable Suburbs" symposium. Even *Time* magazine claimed that "recycling the suburbs" is one of ten ideas that are "changing the world right now" (Walsh 2009). The concept and concern have recently spread beyond the US to the UK and Australia as well (Dixon and Eames 2013).

A number of critical factors make suburban retrofit timely. Many regions now contain a dearth of cheap, undeveloped land with good freeway access. Gas prices and commute times continue to rise. There is also a growing multi-unit market and demand for urban amenities. These factors come together to make the retrofitting of suburban greyfield properties into centralized, mixed-use spaces more attractive. In fact, large suburban residential developers are now creating "urban" divisions to market projects that mimic gentrified urban spaces "offering lofts, yoga studios and billiards lounges" (Dunham-Jones and Williamson 2008: 6) (Figure 8.2).

Suburban retrofit strategies call for revising zoning codes to allow large-scale, multi-use projects. New zoning can help coordinate land uses and encourage the redesign of commercial strips as well. One key strategy is to create central destination points rather than allowing the typical linear shopping strip. Cities can modify their zoning codes to concentrate development in particular intersections on a commercial strip and allow residential use, prohibiting retail in between, as Whittier, CA has done along seven miles of its primary commercial corridor (Tung and Sasaki 2010). Similarly, Atlanta developed a neighborhood commercial zoning category with higher, concentrated densities and pedestrian-oriented street frontages. Alternatively, cities can provide incentives such as streamlining permitting processes or waiving development fees for projects that follow this program. On a smaller scale, zoning changes can also encourage incremental increases in suburban densities by allowing the construction of second units or additions on residential parcels (Pfeiffer 2014, Tachieva 2010).

Additionally, building setback regulations and parking requirements can hinder suburban retrofits. Setback requirements frequently mandate a large space in front of commercial buildings. This discourages pedestrians while encouraging street-facing parking. One alternative is to allow "liner buildings" within the street-front setback of commercial strips (Gamble and LeBlanc 2005). This creates density while reducing the large block size typical of suburbia and thereby encourages walking. At the same time, parking requirements encourage low density because they are typically pegged to the floor area of a building, which can be quite large in suburban environments. Revising parking requirements and seeking alternative temporary uses there can

Figure 8.2   Rockville Town Square in suburban Maryland
Photograph by Federal Realty Investment Trust

serve as potential revitalization schemes (Ben-Joseph 2012). One strategy is to use a shared parking formula. This allows projects to reduce the number of required spaces by meeting parking requirements based on different types of users, such as residents and employees, who will use the same parking spaces at different times of the day.

Whereas these strategies help to adapt the existing suburban stripscape, suburban retrofit projects are also proposed for entire sites. The Belmar Mall retrofit in suburban Lakewood, CO outside of Denver is a good example. This 104-acre mall surrounded by surface parking was transformed into a new 23-block mixed-use area between 2001 and 2012 (Dunham-Jones and Williamson 2008). The retrofit project essentially creates a new downtown containing 1,300 housing units along with office, retail, and civic uses. Many buildings contain ground-floor retail with apartments and offices above. This configuration allows the project to triple the density over the former mall while still incorporating adequate parking and a 2.2-acre park at the center of the site. Buildings are zoned to abut sidewalks within an interconnected street grid (Figure 8.3).

TRANSFORMING THE BOX

Underused or abandoned big box retail is a key focus of suburban retrofit strategies and a critical challenge for suburban revitalization. Given their size, location, and zoning, vacant big box retail is a redevelopment problem particular to suburban areas. When major big box retailers like Walmart leave older locations for newer buildings nearby, they leave behind the older big box store. Increasing online shopping also forced retailers such as Borders Books or Circuit City to downsize or fold. In fact, nearly 35% of the 870.7 million square feet of vacant retail space in the US is in the form of empty big box stores (Schindler 2012). These spaces are not easy to lease or redevelop and, as a result, many communities today are stuck with a growing number of these

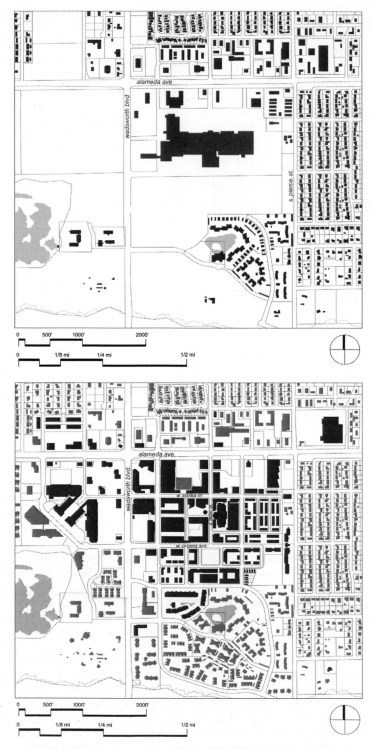

Figure 8.3   Belmar Mall retrofit morphology, 1975 (top) and 2015 (bottom), Lakewood, CO

Diagrams from Dunham-Jones and Williamson (2008), courtesy of Ellen Dunham-Jones and June Williamson

abandoned "ghost boxes" (Christensen 2008, Schindler 2012). One study found that on average the 30 vacant Walmarts in Texas remained vacant for three years (Hunt and Ginder 2005).

The call to transform dying malls and ghost boxes into "urban places" is an important step toward reducing GHG emissions, fighting climate change, and creating community connections. Big box redevelopment projects have included new libraries, recreation centers, schools and community college facilities, religious institutions, museums, and antiques malls (Christensen 2008) (Figure 8.4). They also often provide additional office space and a component of new mixed-use centers. Projects of this nature are important because they help to redirect growth to existing infrastructure rather than continuing expansion into the suburban periphery. They can provide public spaces and new retail options in areas where these are lacking. They also provide an important boost to the municipal tax base that can help fund other improvements.

Figure 8.4   An empty Big Lots store retrofitted into a recreation center. Collinwood Recreation Center, Cleveland, OH
Photograph by Scott Pease, Architect City Architecture

To adapt dead malls and underutilized office parks, proponents of suburban retrofit argue that cities must overhaul suburban zoning codes, which in many municipalities present impediments to retrofitting. Though changing, some suburban zoning ordinances do not allow retail with residential or other uses except in specially designated zones. They also maintain height limits of just two to three stories, which promote low density by preventing the construction of larger multi-story retail or mixed-use projects, and require excessive parking.

Suburban retrofit projects of all types must contend with crucial issues and challenges that other revitalization efforts face. First, most new commercial projects do not create many new jobs. Rather, they shift employment from one area to another rather than providing additional long-term employment in a city or region. Second, market-oriented mechanisms that might help create new "urban" destinations will target wealthier and whiter suburban areas, or the services and goods they provide will be affordable only to wealthier community members. The chance of redeveloping a dead mall in an inner-suburban area struggling with high poverty and unemployment is considerably lower. Unfortunately, these are the areas that most need better access to basic services and employment opportunities. Finally, as advocates of suburban retrofitting recognize, single projects or destinations do not alter driving habits or the broad pattern of urban sprawl, and the majority of areas to be redeveloped are not located on transit lines. As such, we still face the challenge of weaving together these isolated projects throughout a region. More than this, planners cannot focus on design alone, but also must face the reality of suburban diversity and inequality. Given the scale and scope of the issues, achieving suburban revitalization may require larger state and federal initiative. Just as the federal government induced suburbanization during the New Deal and after World War II, perhaps larger regional planning efforts and state-level policy can reverse the trend of low-density suburban sprawl (Box 8.1)?

BOX 8.1 CALIFORNIA'S REGIONAL SOLUTION TO SUBURBAN REVITALIZATION AND CLIMATE CHANGE

Within the past decade, California has passed a pair of bills that have the potential to reconfigure suburban form and spur development that is more environmentally sensitive and socially equitable. Passed in 2006 and initiated in 2012, California Assembly Bill 32 (AB32, or the Global Warming Solutions Act) mandates that greenhouse gas (GHG) emissions statewide be reduced to 1990 levels by 2020 (California Air Resources Board 2014b). One of the Bill's primary outcomes has been the establishment of a cap-and-trade program. The program works by requiring businesses that pollute above a set level or "cap" to purchase or "trade" permits with firms that have not met their emissions cap to enable them to generate more emissions. The intention of a cap-and-trade program is to create an economic incentive to reduce emissions.

However, vehicle emissions are the largest source of pollution in the state, accounting for as much as 30–40% of all GHG emissions. In order to help meet the AB32 mandate, the State also enacted State Bill 375, the Sustainable Communities and Climate Protection Act of 2008. Dubbed the "anti-sprawl bill," the Act aims to reduce emissions through coordinated regional land-use and transportation planning (California Air Resources Board 2014a, Shigley 2009). In conjunction, the regional planning process requires that each of the State's metropolitan planning organizations must develop plans to adequately house the entire regional population. In other words, each region must develop a "sustainable communities strategy" that addresses the environmental issues associated with suburban sprawl not simply by restricting growth but by reconfiguring land-use, transportation, and housing patterns.

To achieve this goal, plans will need to concentrate on strategies to increase the density of traditional suburban land-use patterns. At the same time, they must also distribute multiple housing choices equitably throughout the region and establish transportation options and living and working arrangements that can reduce GHG emissions. The Act also incentivizes local governments and developers to pursue sustainable development by reducing or waiving environmental review requirements and environmental impact reports if the projects meet plan requirements and are developed in desired development zones. Those regions that do not set a sustainable communities plan, or that do not comply with their plan, risk losing their transportation funding.

The Sustainable Communities Bill works from the theory that if we build in close proximity, we can reduce auto dependency and therefore GHG emissions and other environmental problems. In essence, the Bill demands that cities and regions devise strategies to encourage the development of compact, mixed-use places that are accessible by multiple forms of transit by households from different income levels. If successful, the Bill has the potential not only to minimize vehicle pollution, but to provide much-needed transit and housing options to struggling inner-suburban areas.

RESURGENT SUBURBS?

A defining feature of the contemporary inner-ring suburb is its racial and ethnic diversity. Orfield and Luce (2013: 1) go as far as to claim that "suburban communities are now at the cutting edge of racial, ethnic and even political change in America." They point out that there are now twice as many diverse neighborhoods in the suburbs as there are in the central cities. It is important to remember that while 44% of suburbanites in large metros live in racially integrated neighborhoods, many others are situated in neighborhoods segregated by class and race. Like the central cities, the lowest-income areas tend to be segregated communities of color, which lack sustained educational and economic opportunity and experience higher levels of fiscal stress.

Suburban diversity is driven partly by new immigrants who gravitate to the suburbs at higher rates than the city centers. Historically immigrants concentrated in central city neighborhoods, but by the 2010s over half of all foreign-born residents resided in the suburbs of major metropolitan areas such as Los Angeles and New York as well as fast-growing cities such as Dallas and Phoenix. The fastest-growing destinations, however, are cities in the southeast such as Raleigh, NC and Nashville, TN (Frey et al. 2009). According to US Census data, foreign-born residents accounted for one-third of the overall population growth in the suburbs in the first decade of the 2000s. Further, nearly 30% of US immigrants now live in inner-ring suburbs. This means that they are now home to more immigrants and are more diverse than most central cities (Frey et al. 2009, Puentes and Warren 2006, Suro, Wilson, and Singer 2011).

As many inner-suburbs have become the new immigrant gateways (Suro et al. 2011) or "arrival cities" (Saunders 2012), they are faced with a new set of opportunities as well as challenges for urban revitalization. Ethnic immigrant communities are transforming many inner-ring suburbs in complex and often contradictory ways. For many inner-ring suburbs, these communities attract necessary new residents. They also bring new businesses that are staving off suburban decline by boosting local spending and attracting new commercial activity. This improves the local tax base and brings new jobs. These residents and business owners create significant improvements to their physical surroundings through updated storefronts, new signage, landscaping, and home improvements. Not only this, but immigrants tend to stay and invest in their communities rather than moving outward as they become more upwardly mobile. All of these investments occur as an aging baby-boomer population nears retirement (Myers 2007).

At the same time, however, foreign-born suburban residents also experience higher poverty rates than US-born citizens and face language, health care, and education barriers (Suro et al. 2011). While not uniformly the case, they are also more likely to live in economically distressed neighborhoods, experience housing and cultural discrimination and, as a result, live in segregated neighborhoods (Orfield and Luce 2013). Consequently, while older, inner-ring suburbs in particular stand to benefit from their growing foreign-born populations, they also face challenges related to economic opportunity, housing, and social service provision.

Many suburban areas have only recently become home to substantial immigrant communities and do not have the history and experience to adapt to the changing demographics. A few cities, such as Dayton, OH, recognize the potential of their immigrant communities to spur revitalization and have plans in place to attract and support them. Dayton, for example, has set up programs to tutor foreign students, assist immigrant entrepreneurs, translate documents, and has instructed police not to check the immigration status of people who report crimes or are suspected of minor offences (Altman 2014). Other places, however, fear change. Suburban cities such as Farmers Branch, TX and Hazelton, PA recoil from the reality of changing demographics and have enacted laws intended to prevent those who entered the country illegally from living or obtaining services within their borders (Vicino 2013, Vitiello 2013). Even when these laws are struck down as unconstitutional, the attempt sends a message to people of color that they are not welcome.

Despite the challenges, inner-ring suburbs benefit from and can support ethnic immigrant communities in a number of ways. For one, these communities bring a new crop of entrepreneurs and businesses that can stimulate local economic development. According to the protected market hypothesis, ethnic entrepreneurs are able to establish a niche within their community by tapping into ethnic social and business networks and their knowledge of community preferences (Light 1972). These enterprises can provide new jobs and often employ co-ethnics and individuals from other minority communities who may otherwise face restricted opportunities in the mainstream labor market due to their legal status and language barriers. Their enterprises span an array of sectors from professional and educational services to restaurant and retail, and immigrant entrepreneurs often provide services in niche and underserved markets. This helps to diversify and fill gaps in the local economy while capturing tax revenues (Bowles and Colton 2007, Liu, Miller, and Wang 2014).

Immigrant entrepreneurs, businesses, and nonprofit organizations also play a key role in remaking the built environment. In contrast to the suburban retrofit agenda, which treats the suburban commercial strip as a planning problem or an eyesore to be rebuilt, small business owners and community organizations seek out the aging, low-slung commercial strips as the raw material for their enterprises.

As Davis (1997) points out, like Jane Jacobs, immigrant entrepreneurs see the importance in old, ugly, and ordinary buildings and bring a vitality to them that was formerly absent. While certainly lacking the urbane qualities of Jacobs' Greenwich Village, the inner-suburban commercial strip likewise provides affordable, easily adaptable space within close proximity to customers and community members. In this way, the presence of aging commercial strips and vacant big boxes provides an important incubator that has spurred the revitalization of many inner-suburban communities. Little Taipei in suburban Monterey Park outside Los Angeles, Little Saigon in Orange County, CA, and Buford Highway in Atlanta have all undergone significant economic and physical change at the hand of ethnic entrepreneurs. New organizations and businesses here have transformed underused or dilapidated areas into regional shopping destinations and even tourist magnets by bringing new restaurants, grocery stores, retail services, health clinics, job centers, and places of worship (Box 8.2, Figure 8.5). Surrounding cities have benefited tremendously from the injection of spending by consumers attracted by specific ethnic goods and services, improvements to street façades and commercial buildings, and improvements to public safety and spaces (Davis 1997, Liu et al. 2014, Loukaitou-Sideris 2000).

Suburban municipalities can foster this development. For example, Chamblee, GA on the Buford Highway enacted an overlay zoning district that establishes building design standards

Figure 8.5 Little Saigon shopping center, Westminster, CA
Photograph by Carl Grodach

(e.g. use of colorful façades and decorative storefronts), requires multilingual signage, and encourages the creation of pedestrian amenities and public spaces (Liu et al. 2014). Cities may also encourage the reuse of vacant big box stores and large retail buildings around the needs and preferences of surrounding ethnic communities by targeting redevelopment incentives and small business assistance toward community-specific uses. For example, the City of Minneapolis and its development partners helped to establish an international marketplace on the ground floor of a redeveloped Sears building that contains grocers and food vendors from the diverse local immigrant population (Carr and Servon 2008).

Often ethnic and immigrant entrepreneurs also form merchants' associations to represent their interests (Loukaitou-Sideris 2000, Sutton 2010). For example, the Pico-Union Merchants Association brought together an array of business owners to collectively address common problems such as crime and sanitation. They engaged in area promotion, marketing the area as the Byzantine-Latino Quarter, and lobbied the City of Los Angeles for streetscape and façade improvement grants (Loukaitou-Sideris 2000). These entities, as well as other nonprofit community associations, can also be central in delivering crucial public services, providing legal information and meeting spaces, and sponsoring public festivals. This in turn functions to extend social ties within the community and potentially can help to overcome language and cultural barriers.

At the same time that they provide important benefits, concentrated ethnic communities also present challenges to suburban municipalities. The security and familiarity that immigrant communities provide may contribute to isolating members from the larger community, thus slowing down their ability to navigate the wider labor market and engage in local business circles. Cities can help to address these issues by establishing social service centers that provide counseling and access to various resources. They can engage in partnerships with community colleges to provide targeted language and business classes. They can also follow the lead of cities like New York City and organize small business assistance and lending programs targeted to specific immigrant communities.

The diversity of immigrants and the varied economies and politics of suburban communities combine to produce a complex and conflicting set of opportunities and challenges for suburban communities. In the years to come, the growing diversity and establishment of inner-ring suburban immigrant communities will play an important role in shaping revitalization strategies and outcomes.

BOX 8.2 LITTLE SAIGON: THE EVOLUTION AND PLANNING CHALLENGES OF A MATURE SUBURBAN ETHNIC ENCLAVE

The suburban commercial district of Little Saigon in Orange County, CA is the oldest and largest concentration of Vietnamese Americans and Vietnamese American-owned businesses in the US. The district emerged in the 1970s to serve the Vietnamese refugees and immigrants settling in and around the majority white and politically conservative city of Westminster. As a second wave of ethnic Chinese immigrants from Vietnam became established in the area in the 1980s, the business district grew rapidly into a Southeast Asian cultural destination, attracting hundreds of thousands of tourists and serving as an important source of cultural connection and community for its residents. Since its origins in the 1970s, Little Saigon has evolved into an economic hub that contains 4,000 Vietnamese American-owned businesses (Orange County Register 2014) and is "one of Orange County's most visible and influential cultural exports. The densely packed neighborhood of shopping malls and phở joints generates trends in music, food and business that resonate around the world—even in Vietnam itself" (Hinch 2012).

Little Saigon grew initially because it served as a community center for immigrants attracted to the area by its relatively affordable housing and job opportunities in the aerospace and defense industries. The area provided social services for the Vietnamese community and became a commercial center that included markets, restaurants, karaoke bars, music and video stores, and specialized crafts such as tailors, traditional dressmakers, and jewelers, which primarily served the Vietnamese residents (Aguilar-San Juan 2009, Lieu 2011). The retail agglomeration and social support system enabled entrepreneurs to start their own small businesses and sustained the rise of traditional Vietnamese culture that began to remake the suburban enclave. In 1988, as the Little Saigon economy began to boom and the area contained a greater concentration of Asian residents, it received official designation from the State of California. Still, Vietnamese and Chinese residents faced discrimination. A petition was circulated—and defeated—to bar Vietnamese immigrants from acquiring business licenses, real estate red lining occurred, and vandals defaced "Little Saigon" signs (Lieu 2011).

As the community came of age, it evolved into a larger commercial center that marketed ethnic identity to tourists and the community. The 1980s saw the construction of a pagoda and other markers of Chinese and Vietnamese culture. Additionally, an ethnic Chinese investor developed a series of large shopping malls targeted at the growing Southeast Asian population in Orange County. Hundreds of small shops thrived inside the themed Asian-style architecture of the malls. The City of Westminster supported the architectural character by developing design standards for the area in 1993. The overlay zone specified architectural treatments intended to promote ethnic identity by requiring that all new commercial buildings "incorporate architectural elements similar to those found on buildings constructed in Vietnam in the early 1900s in the French Colonial tradition" or a "traditional Chinese architectural theme" (Lieu 2011: 48). The Vietnamese Chamber of Commerce also promoted tourism through guided tours for investors and trade missions to Southeast Asia (Aguilar-San Juan 2009).

Some argue that the architecture, landscaping, businesses, and community services carved out a unique sense of place in a homogenous suburban landscape. Others, including many Vietnamese residents view the symbols of colonialism and Chinese heritage as inauthentic ethnic theming that create interethnic conflict within the district's business community (Aguilar-San Juan 2009).

Recent years have seen local business owners and city officials concerned with sustaining Little Saigon as a commercial and cultural center. According to a study by the Urban Land Institute (2007), some business and property owners who do not speak English habitually violate building and zoning codes, and others do not pay state sales tax. Existing land-use patterns and zoning are still geared toward low-density suburban development and the area does not contain a distinct center or gateways. Little Saigon also struggles with traffic, parking, and pedestrian safety in its dense sprawl of buildings. Finally, informants critique the architectural standards as "not authentically Vietnamese." To address these issues, the City of Westminster recently rezoned a main street in the district from light industry to mixed use and has developed plans to encourage more pedestrian activity and promote the development of a mixed-use center in Little Saigon (Orange County Register 2014).

SUMMARY

Suburban decline is an increasingly urgent, but sometimes overlooked, facet of urban revitalization. As the suburbs have expanded, aged, and become more diverse, some areas have begun to experience growing pains traditionally associated with the central city. Some, pointing to region-wide issues related to poverty and class and race-based segregation in inner-ring suburbs, call for federal and state-level solutions. Others seek to redesign and retrofit suburbia to be more socially and environmentally sustainable. Still others look to the growing foreign-born population as a source of economic and physical revitalization. Each concern represents an important challenge that cities will have to address in the near term in order to create an equitable and sustainable environment.

STUDY QUESTIONS

1   What issues distinguish approaches to suburban revitalization from those in central city neighborhoods? Recognizing that the suburbs form a diversity of places, can you identify an overarching set of priorities for suburban revitalization?
2   Poverty is a growing challenge for many inner-suburbs. Drawing on the various strategies for suburban revitalization in this chapter, what do you think would be most effective in alleviating suburban poverty?
3   Using available information for real estate values and rental property affordability, identify inner-ring suburbs that appear to be thriving and those that might be suffering from disinvestment and decline. Visit one of each and describe the conditions. Did you see evidence of decline? Did you see vibrant areas? Did you see opportunities to redesign the physical environment? What would you prioritize?
4   In what ways are suburban retrofit strategies and those that support emerging ethnic entrepreneurs and suburban commercial districts at odds with each other? In what ways could suburban municipalities incorporate these two approaches to spur revitalization?

NOTE

1   See Grant (2006) for a more detailed description of new urbanism.

REFERENCES

Aguilar-San Juan, Karin. 2009. *Little Saigons: Staying Vietnamese in America*. Minneapolis: University of Minnesota Press.

Altman, A. 2014. "One Ohio city's growth strategy? Immigrants." *Time* June 5. http://time.com/2826298/one-ohio-citys-growth-strategy-immigrants

Ben-Joseph, E. 2012. *Rethinking a lot: the design and culture of parking*. Cambridge, MA: MIT Press.

Bowles, J and Colton, T. 2007. *A world of opportunity*. New York: Center for an Urban Future.

California Air Resources Board. 2014a. "Assembly Bill 32 overview." Sacramento, CA: California Air Resources Board. www.arb.ca.gov/cc/ab32/ab32.htm

California Air Resources Board. 2014b. "Sustainable communities." Sacramento, CA: California Air Resources Board. www.arb.ca.gov/cc/sb375/sb375.htm

Carr, JH and Servon, LJ. 2008. "Vernacular culture and urban economic development: thinking outside the (big) box." *Journal of the American Planning Association* 75(1): 28–40. doi: 10.1080/01944360802539226

Christensen, J. 2008. *Big box reuse*. Cambridge, MA: MIT Press.

Davis, T. 1997. "The miracle mile revisited: recycling, renovation, and simulation along the commercial strip." *Perspectives in Vernacular Architecture* 7: 93–114.

Dixon, T and Eames, M. 2013. "Urban retrofitting for the transition to sustainability." *Building Research and Information* 41(5).

Duany, A, Plater-Zyberk, E, and Speck, J. 2010. *Suburban nation: the rise of sprawl and the decline of the American Dream*. New York: North Point Press.

Dunham-Jones, E and Williamson, J. 2008. *Retrofitting suburbia: urban design solutions for redesigning cities*. Hoboken, NJ: John Wiley.

Ewing, R and Hamidi, S. 2014. *Measuring Sprawl 2014*. Washington, DC: Smarth Growth America.

Forsyth, A. 2012. "Defining suburbs." *Journal of Planning Literature* 27(3): 270–281. doi: 10.1177/0885412212448101

Frey, WH, Berube, A, Singer, A, and Wilson, JH. 2009. *Getting current: recent demographic trends in metropolitan America*. Washington, DC: Brookings Institution.

Gamble, M and LeBlanc, J. 2005. "Incremental urbanism." *Harvard Design Review* March 17. www.planetizen.com/node/15797

Grant, J. 2006. *Planning the good community: new urbanism in theory and practice*. Royal Town Planning Institute Library Series Vol. 9. London: Taylor & Francis.

Grant, JL, Nelson, AC, Forsyth, A, Thompson-Fawcetts, M, Blais, P, and Filion, P. 2013. "Suburbs in transition." *Planning Theory and Practice* 14(3): 391–415. doi: 10.1080/14649357.2013.808833

Green Leigh, N and Lee, S. 2005. "Philadelphia's space in between: inner-ring suburb evolution." *Opolis* 1(1): 13–32.

Hanlon, B. 2009. *Once the American Dream: inner-ring suburbs of the metropolitan United States*. Philadelphia, PA: Temple University Press.

Hanlon, B, Vicino, T, and Short, JR. 2006. "The new metropolitan reality in the US: rethinking the traditional model." *Urban Studies* 43(12): 2129–2143.

Hayden, D. 1997. *The power of place: urban landscapes as public history*. Cambridge, MA: MIT Press.

Hinch, Jim. 2012. "O.C.'s Saigon? Nothing little about it." *Orange County Register* November 4. www.ocregister.com/articles/vietnamese-376486-saigon-little.html

Hunt, HD and Ginder, J. 2005. "Lights out: when Wal-Marts go dark." *Tierra Grande* April.

Jacobs, J. 1961. *The death and life of great American cities*. New York: Random House.

Kneebone, E and Berube, A. 2013. *Confronting suburban poverty in America*. Washington, DC: Brookings Institution.

Kunstler, JH. 1994. *Geography of nowhere: the rise and decline of America's man-made landscape*. New York: Simon & Schuster.

Lee, S and Green Leigh, N. 2007. "Intrametropolitan spatial differentiation and decline of inner-ring suburbs: a comparison of four US metropolitan areas." *Journal of Planning Education and Research* 27(2): 146–164. doi: 10.1177/0739456X07306393

Lieu, Nhi T. 2011. *The American Dream in Vietnamese*. Minneapolis: University of Minnesota Press.

Light, IH. 1972. *Ethnic enterprise in America: business and welfare among Chinese, Japanese, and Blacks*. Ewing, NJ: University of California Press.

Liu, CY, Miller, J, and Wang, Q. 2014. "Ethnic enterprises and community development." *GeoJournal* 79, 565–576. doi: 10.1007/s10708-013-9513-y

Loukaitou-Sideris, A. 2000. "Revisiting inner-city strips: a framework for community and economic development." *Economic Development Quarterly* 14(2): 165–181.

Lucy, W and Phillips, D. 2000. *Confronting suburban decline: strategic planning for metropolitan renewal.* Washington, DC: Island Press.

Marshall, WE, Piatkowski, DP, and Garrick, NW. 2014. "Community design, street networks, and public health." *Journal of Transport and Health* 1(4). doi: 10.1016/j.jth.2014.06.002

Myers, D. 2007. *Immigrants and boomers: forging a new social contract for the future of America.* New York: Russell Sage Foundation.

Orange County Register. 2014. "The heart of Westminster: one city block." *Orange County Register* February 21. www.ocregister.com/articles/home-602771-one-city.html

Orfield, M. 1997. *Metropolitics: a regional agenda for community and stability.* Washington, DC: Brookings Institution Press.

Orfield, M and Luce, TF. 2013. "America's racially diverse suburbs: opportunities and challenges." *Housing Policy Debate* 23(2): 395–430.

Pfeiffer, D. 2014. "Retrofitting suburbia through second units: lessons from the Phoenix region." *Journal of Urbanism* 8(3): 279–301. doi: 10.1080/17549175.2014.908787

Puentes, R and Orfield, M. 2002. *Valuing America's first suburbs: a policy agenda for older suburbs in the Midwest.* Washington, DC: Brookings Institution Center on Urban and Metropolitan Policy.

Puentes, R and Warren, D. 2006. *One-fifth of America: a comprehensive guide to America's first suburbs.* Washington, DC: Brookings Institution Center on Urban and Metropolitan Policy.

Randolph, B and Tice, A. 2014. "Suburbanizing disadvantage in Australian cities: sociospatial change in an era of neoliberalism." *Journal of Urban Affairs* 36(s1): 384–399. doi: 10.1111/juaf.12108

Saunders, D. 2012. *Arrival city: how the largest migration in history is reshaping our world.* New York: Vintage Books.

Schafran, A. 2013. "Discourse and dystopia, American style: the rise of 'slumburbia' in a time of crisis." *City* 17(2): 130–148.

Schindler, S. 2012. "The future of abandoned big box stores: legal solutions to the legacies of poor planning decisions." *University of Colorado Law Review* 83: 471–548.

Shigley, Paul. 2009. "California's aerial combat: the state tries a first-in-the-nation approach to attacking climate change." *Planning* February: 11–15.

Short, JR, Hanlon, B, and Vicino, TJ. 2007. "The decline of inner suburbs: the new suburban gothic in the United States." *Geography Compass* 1(3): 641–656.

Suro, R, Wilson, JH, and Singer, A. 2011. *Immigration and poverty in America's suburbs.* Washington, DC: Metropolitan Policy Program, Brookings Institution.

Sutton, SA. 2010. "Rethinking commercial revitalization: a neighborhood small business perspective." *Economic Development Quarterly* 24(4): 352–371.

Tachieva, G. 2010. *Sprawl repair manual.* Washington, DC: Island Press.

Talen, E. 2011. "Sprawl retrofit: sustainable urban form in unsustainable places." *Environment and Planning B* 38(6): 952.

Tung, F and Sasaki. 2010. *Restructuring the commercial strip: a practical guide for planning the revitalization of deteriorating strip corridors.* United States Environmental Protection Agency.

Urban Land Institute. 2007. *Little Saigon, Westminster, California.* Washington, DC: Urban Land Institute.

Vicino, TJ. 2008. "The quest to confront suburban decline political realities and lessons." *Urban Affairs Review* 43(4): 553–581.

Vicino, TJ. 2013. *Suburban crossroads: The fight for local control of immigration policy.* Lanham, MD: Lexington Books.

Vitiello, D. 2013. "The politics of immigration and suburban revitalization: divergent responses in adjacent Pennsylvania towns." *Journal of Urban Affairs* 36(3): 519–533.

Walsh, B. 2009. "10 ideas changing the world right now." *Time* March 12. http://content.time.com/time/specials/packages/article/0,28804,1884779_1884782_1884756,00.html

Williamson, J. 2013. *Designing suburban futures: new models from build a better burb.* Washington, DC: Island Press.

9

Re-envisioning shrinking cities

INTRODUCTION

Some cities have lost overall population decade after decade. This has led to problems such as high levels of vacant land, declining infrastructure, and fewer opportunities for residents who stay. The city of Detroit is the iconic US example as it fell from a peak population of close to 2 million in 1950 to less than 800,000 in 2010. Nevertheless, Detroit is not alone. Between 1950 and 1980, 24 of the 50 largest US cities lost population, and from 1980 to 2000, 12 of the country's largest 50 cities lost population (Beauregard 2009). Table 9.1 shows population loss in ten US cities since 1950. In 1950, all of these cities ranked in the list of the 20 largest US cities.

Population shifts reflect uneven development both within and among regions. In many US cities, population loss can be attributed to two interrelated trends that we discussed in chapters 2 and 4: deindustrialization and suburbanization. Many depopulating cities and regions that had strong manufacturing economies have not easily transitioned to new areas of economic growth. In addition, fragmented regional governments prevent regional collaboration, resulting in the primary city losing population even as the surrounding suburbs grow (Figure 9.1). Regional racial politics in the US have been a particularly destructive dimension of this. Housing discrimination has created regional landscapes where some jurisdictions are wealthier and have a higher proportion of white residents. These suburbs do not consider it advantageous to collaborate or see their interests entwined with the primary city, inhibiting effective regional action (Thomas 2013). Because population shifts result from relative opportunities that develop in one region when compared with another, cities and regions can lose population in fast- and slow-growing countries alike. The effect of population loss is widespread property abandonment, declining municipal revenues and household opportunities, and infrastructure that is outsized for the current population.

Can we adapt shrinking cities to enable residents who stay to thrive and prosper? If so, how can we ensure that these adaptations meet all objectives of robust revitalization, including the provision of opportunities to build human capital, increase social–cultural equity, improve the built environment, and create jobs for all while improving environmental quality? In this chapter we discuss the particular circumstances and opportunities for shrinking cities. The circumstances in cities that lose population differ in intensity from those in growing cities, and the responses reflect the unique conditions. Residents and their allies take action that ranges from maintaining empty lots to building art installations. Public agencies have developed strategic plans and land

TABLE 9.1 POPULATION LOSS IN 10 US CITIES SINCE 1950

City	1950	1960	1970	1980	1990	2000	2010	Decline (%)
Philadelphia	2,071,605	2,002,512	1,948,609	1,688,210	1,585,577	1,517,550	1,526,006	26
Detroit	1,849,568	1,670,144	1,511,482	1,203,339	1,027,974	951,270	713,777	61
Baltimore	949,708	939,024	905,759	786,775	736,014	651,154	620,961	35
Cleveland	914,808	876,050	750,903	573,822	505,616	478,403	396,815	57
St. Louis	856,796	750,026	622,236	453,085	396,685	348,189	319,294	63
Pittsburgh	676,806	604,332	520,117	423,938	369,879	334,563	305,704	55
Milwaukee	637,392	741,324	717,099	636,212	628,088	596,974	594,833	20
Buffalo	580,132	532,759	462,768	357,870	328,123	292,648	261,310	55
New Orleans	570,445	627,525	593,471	557,515	496,938	484,674	343,829	45
Cincinnati	503,998	502,550	452,524	385,457	364,040	331,285	296,943	41

Source: US Census, Selected Historic Census Data 1790–1990; US Census Population Change and Distribution: 1990–2000; US 2010 Census

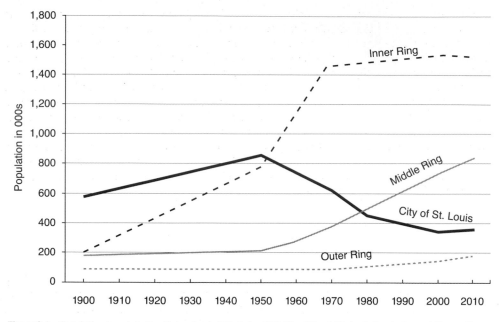

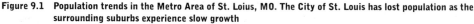

Figure 9.1   Population trends in the Metro Area of St. Loius, MO. The City of St. Louis has lost population as the surrounding suburbs experience slow growth

Courtesy of Wendell Cox

banks to manage vacant property. Increasingly, residents along with public officials, architects, and planners have re-envisioned shrinking cities around land-intensive conversions such as urban agriculture and lower-density settlement patterns with focused nodes of redevelopment. Nevertheless, the challenges are great. In the next section, we highlight the debate over planning to shrink. The following section explains the major challenges facing shrinking cities. We then turn to myriad responses from the public sector and neighborhood residents as they attempt to adapt the built environment to a smaller population in ways that results in stable, livable cities. We finally return to the question of growth and discuss how local governments in shrinking cities plan for growth and adapt to a smaller population.

SHOULD WE PLAN FOR URBAN SHRINKAGE?

While it is clear that some cities and regions have sustained population loss for decades, how to revitalize these areas continues to be debated. Cities and regions such as Buffalo and Pittsburgh in the US, Manchester in the UK, or the Ruhr Valley in Germany experienced massive job losses and population outmigration, and these cities and regions have struggled to varying degrees to find their role in a restructured, global economy. It has been easier to identify the causes of population loss than to design effective solutions, however.

To address the challenges, some began to think about urban shrinkage as a distinct process that requires new approaches for effective action. The term "shrinking cities" emerged from a three-year Shrinking Cities initiative funded by Germany's Federal Cultural Foundation (Shrinking Cities n.d.). The initiative sought to examine four shrinking cities and regions internationally to begin to re-envision these places with temporary, flexible uses as well as new ways to use unoccupied buildings and vacant land. Shrinking cities as a concept has gained currency and been widely adopted (Bontje 2005, Hollander et al. 2009, Hollander and Németh 2011, Martinez-Fernandez et al. 2012, Richardson and Nam 2014, Ryan 2012).

From some perspectives, the term shrinking cities has negative connotations. In the US, some observers prefer the term "legacy cities" to specifically address the challenges facing former industrial cities in the northeast (Mallach and Brachman 2013). Some preservationists also prefer this term because it draws attention to the historical importance of the city's architecture and urban spaces. In other cases, people simply refer to cities growing smaller (Rugare and Schwarz 2008).

Because the regions overall often are not losing population, another perspective focuses on regional approaches rather than accepting population loss. Since urban processes are dynamic, many people work to stimulate growth rather than plan for shrinkage. Opponents who discourage labeling cities as shrinking argue that disinvestment creates the conditions for reinvestment, and argue that defining a region as "shrinking" appears to be acknowledging failure. Whatever they are called, depopulating cities face real changes. We address the question of growth at the end of this chapter, but first explain the unique challenges cities face when they lose population and the myriad creative responses from people living and working in these cities.

THE UNIQUE CHALLENGES FACING CITIES AS THEY GROW SMALLER

In some ways, neighborhoods in shrinking cities face similar circumstances to other areas seeking revitalization. Residents are concerned about blighted property. They need jobs and opportunities, parks, and other urban amenities. Because established industries have downsized or left these regions, many people who earned higher wages moved away. As a result, median household incomes have declined and residents who stay have fewer opportunities for upward mobility. In all cases, residents work to improve their neighborhood conditions. Nevertheless, shrinking cities create particularly challenging circumstances because overall demand for land and services is low, which results in vacant property and financial stress.

CHALLENGE 1: VACANT AND ABANDONED PROPERTY

Vacant and abandoned property is the most visible challenge for shrinking cities. Vacant properties are a normal component of real estate markets as residents and businesses move around and new development comes online. In shrinking cities, however, overall demand for property is on the decline. For example, in 2012, 12% or 43,680 parcels in New Orleans were estimated to be unoccupied (vacant lots, blighted structures, or vacant but habitable residences) (Plyer and Ortiz 2012). Detroit has approximately 78,500 vacant structures, a fifth of its total housing stock, and 114,000 empty parcels (*New York Times* 2014). Cities must also address commercial vacancy, which is higher in shrinking cities than in the country as a whole. For instance, in 2012 Detroit had 29,276 vacant structures or 41% vacancy in commercial properties compared with New York, which also had high commercial vacancy at 35,869 structures or 26% (Eppig and Brachman 2014).

With too few tenants and buyers, property owners cannot lease or sell their property and they lose incentive to maintain their property. As values drop, property owners become unable to get home improvement loans and, as rents decrease, rental property owners have little reason to make repairs. As property deteriorates, or when owners cannot find tenants, they may abandon the property rather than paying taxes and continuing upkeep. As more property owners stop paying taxes and ignore citations to secure or repair dangerous vacant property, more property is abandoned (Figure 9.2). This creates a downward spiral that damages neighborhood quality (Keenan, Lowe, and Spencer 1999, Accordino and Johnson 2000, Dewar and Thomas 2013, Virginia Tech Metropolitan Institute 2006). In addition, poor conditions of existing buildings, complicated property title problems when property has been empty for a long period, and property owners who are hard to find make redevelopment more difficult.

Figure 9.2 Vacant property and empty lots are more common than occupied property in this formerly dense neighborhood in Detroit, MI

Photograph by Marla Nelson

Vacant and abandoned property impacts local real estate markets and property dynamics. It depresses other property values in the area. In Philadelphia, one analysis showed that properties within 150 feet of vacant property lost $7,000 in value. Vacant properties are also dangerous. According to the US Fire Administration, each year over 12,000 fires break out in vacant properties as a result of arson, carelessness, or faulty wiring (Bass et al. 2005).

Vacant and abandoned property also has direct, adverse impacts on residents' sense of well-being. Residents identify many problems with living near abandoned property. They worry about issues ranging from garbage dumping and feral animals to neighborhood stigma and fear of crime, drug use, and other potentially dangerous activities. Residents report feeling helpless, sad, and depressed when asked about vacant land in their neighborhoods, and they state that abandoned property creates divisions among neighbors and contributes to a lack of community cohesion (Garvin et al. 2013).

In many cities, central neighborhoods have higher levels of abandoned property—an ongoing legacy of the housing programs and white flight discussed in earlier chapters. While this is also true in shrinking cities, the number of neighborhoods in distress is much higher, with the majority of neighborhoods experiencing population loss, deteriorating housing and infrastructure, and abandoned commercial properties. In Detroit, the most extreme example, 99% of census tracts are impacted. Fifty three percent of all Pittsburgh census tracts show some signs of distress and 35% are moderately to severely distressed (Hackworth 2014). Increasingly, cities such as Buffalo also find that their inner-ring suburbs are experiencing similar problems (Virginia Tech Metropolitan Institute 2006).

CHALLENGE 2: OPERATING IN A CONTEXT OF RAPIDLY DECLINING RESOURCES

Severe budgetary constraints are another great challenge facing shrinking cities. As cities lose population, revenues decline. Local sales tax revenues decrease as fewer people live and work in the city and purchase goods and services locally. As fewer people purchase fewer goods and services, businesses including retail establishments close or relocate to more populous areas, further reducing sales tax revenues. At the same time, as businesses leave, the remaining residents have fewer local options and travel farther—and often to surrounding jurisdictions—to shop. As people leave, property values also decrease, reducing property tax revenues. As more property owners abandon their property and stop paying taxes, revenues decline further. In addition, cities have ongoing pension obligations that were built on the assumption that the cities would grow or at the very least retain a high population, resulting in pension obligations that strap the city's finances.

The costs of public services including municipal water and sewage systems, transit systems, and trash collection are partially offset by user fees. Because the number of users declines, the revenues that help cover operating expenses also decrease. In shrinking cities, people continue to live throughout the city even if fewer people live in each neighborhood, and utilities and services have to cover the same land area as they did when the population was larger. The water pipes and street lights must be maintained and police and fire services must be available even if only two houses on a block are occupied. Garbage trucks pass each parcel even if only one house is occupied. Although cities can raise property or sales taxes, steep taxes in areas with poor services create a disincentive to live in the city, encouraging people to "vote with their feet" and move to nearby jurisdictions. The situation can become so dire that cities may declare bankruptcy, as the city of Detroit did in 2013.

CHALLENGE 3: DEMOLITION AND PRESERVATION

A related challenge is determining how to strategically demolish unwanted property. It is easier and less expensive to maintain a vacant lot than keep an empty house secured and minimally maintained. Empty houses are threatening to local residents, who may fear that the house could be taken over by undesirable activities or catch fire. Abandoned buildings also bring down property values more than renovated houses or well-maintained vacant lots (Bass et al. 2005). Nevertheless, it is also costly to demolish property, and cities do not have the authority to demolish property just because it stands empty.

Another complicating factor is the quality and significance of the structures in empty neighborhoods. The central cities' emptiest neighborhoods are often defined by their historic housing stock, empty warehouses, and other industrial buildings. While demolishing structures is necessary for the residents' well-being, preservationists fear that demolishing unoccupied properties will be an irreparable loss of historic architecture and destroy the historic urban fabric that makes the city unique. Some historic neighborhoods in shrinking cities have experienced revitalization, and warehouses turned into loft condominiums have some of the highest property values in Cleveland, Detroit, and St. Louis.

Decisions about preservation are not always simple, however. The abandoned rail stations in formerly thriving industrial cities such as Buffalo, Cincinnati, and Detroit show the tension. Buffalo's outstanding Art Deco Central Terminal stands empty in 2015. Opening in the 1920s, it remained a busy station through World War II, designed to handle over 200 trains and 10,000 passengers daily. By 1955, its use had declined and although New York Central Rail tried to sell it, at that time there was no buyer for such a large building built for such a specific use. It continued to be partially used until 1979, when it stopped operating altogether (Central Terminal Restoration Corporation n.d.). Although the Terminal was purchased by a preservationist in 1997 for $1 and unpaid taxes, it has yet to be restored and put back into use. In 2015, the Central Terminal Restoration Corporation is attempting to raise $65 million to restore the over 523,000 square foot building on a 17-acre lot. It is an uphill battle, however, because it is unclear what new tenants it could attract.

There are many examples of successful adaptive industrial reuse in shrinking cities, including Larkinville in Buffalo, where an entertainment destination was constructed by the former Larkin

Soap Factory (Kurzius 2014, Central Terminal Restoration Corporation n.d.). The sheer number of industrial buildings and their poor condition, however, makes restoring all of them difficult. In Detroit, the situation is so dire that ruin tourism has become normal and visitors come to see the decline. The ruins have been depicted in numerous photography books, including Yves Marchand's and Romain Meffre's (2010) *The Ruins of Detroit*, and in museum exhibits.

RESPONSES AND OPPORTUNITIES

Residents and public officials have responded to these challenges with creativity and innovation. How to maintain vacant land and control blighted properties has been a critical focus, and the public sector and residents both seek to manage local changes. In the following subsections, we first focus on public sector actions to acquire abandoned property. We next discuss residents' responses to increased vacancies and empty parcels. Many people see the opportunity to use these for more land-intensive uses including urban farms and orchards, while some residents respond in other creative ways to the changes they observe, and we examine these actions.

ACQUIRING AND MANAGING VACANT PROPERTY

To manage the apparently insurmountable challenge of abandoned properties, the first tool that cities have is their building codes and code enforcement. Cities have building codes that include basic standards for maintaining and securing property and they can cite offending property owners when properties are not compliant with the standards set forth in the city code. Occupied housing can have code violations if holes develop in the roof or walls that allow water or wind inside, if plumbing or electricals deteriorate causing health or safety problems, or even if the exterior becomes extremely run down. When properties stand empty, they must be maintained or they will become in violation of city codes. For example, in New Orleans grass over 18 inches must be cut and empty properties must be secured. When properties are abandoned, over time the windows break and the lots get overgrown. This can also happen if a property owner moves from the area and the house is not leased. While cities have the authority to improve the appearance of properties by securing them and cutting the grass, and they can assess costs to the property owners, it is difficult for cities under fiscal stress to maintain private properties on a large scale.

To ensure compliance with city standards, cities begin with citing the offending property owners when properties are not compliant with the standards as outlined in the city code. After being cited, if the property owner abates the conditions, the city will often waive part of the assessed fines. If the property owners do not abate the nuisance and pay their fines, the fines will become a lien on the property. However, if the property owner has no use for the property (for example, there is no-one willing to rent or lease the property), and has no way to dispose of the property (for example, there is no-one interested in buying the property), the property owner might decide it is not worth paying the fines or paying to maintain the property. For similar reasons, property owners might stop paying property taxes. Unpaid taxes can also become a lien. In these cases, the property owner becomes willing to relinquish the property. If a city or county has a land bank, a tool discussed later, that will accept unwanted property, this can prevent a lengthy and expensive code enforcement process. Otherwise, the property will fall into further disrepair before being transferred to public ownership.

When properties have a lien, cities and counties have a process whereby they can acquire the properties. The city or county can go through a foreclosure process to recoup the money it owed or acquire the property in exchange for the owed fees. Foreclosure refers to the process by which the city recoups the fines through selling the property. It is often also referred to as a tax sale. In most cases, properties are first put up for auction. The buyer is responsible for paying unpaid fines or taxes. Those properties that are not sold at auction remain in public ownership. In some circumstances, local governments can also use eminent domain to acquire property.

Acquiring the land is only one step. Cities must then decide what to do with the land they own. It is difficult for shrinking cities under fiscal stress to maintain a large inventory of properties. All courses of action have immediate and long-term costs. There are four main objectives when cities acquire vacant and abandoned property:

- hold and maintain properties for which there is no immediate use
- return property to productive commercial, nonprofit, or residential use where demand exists
- facilitate transfer to adjoining properties where residents remain to maintain or use the property
- transform the property to a less intensive use (such as a park, greenway, or garden) where demand for active use is low.

Particularly because there is limited demand for land in shrinking cities, cities and counties are increasingly considering land banks as a crucial tool for long-term property management. Land banks are public or nonprofit authorities that can acquire, manage, hold, and strategically dispose of surplus property (Alexander 2011). Although land banks were first proposed both to manage vacant urban land in weak markets or slow-growing areas *and* to acquire land in fast-growing areas for future public uses, they have been used primarily to address vacant land in areas with high levels of abandonment.

The first US land banks were established in the 1970s in St. Louis and Cleveland. In 1989 Louisville established a land bank and, in 1991, Atlanta followed suit. These early land banks were successful, given their limited powers, but their ability to acquire and manage property was constrained.

A second generation of land banks began in Michigan. In 1999 and 2003, the state enacted significant legislative reforms that made it easier for public authorities to acquire tax-delinquent property. In 2002, the Genesee County Land Bank (beginning as the Genesee County Land Reutilization Council, Inc.) was established in collaboration with the city of Flint. Soon after the Michigan legislation allowed Michigan counties and the city of Detroit to form land bank authorities and, as of 2011, 35 counties and the city of Detroit have established land banks. In 2008 and 2010, Ohio enacted parallel reforms and the Cleveland Land Bank was succeeded by the Cuyahoga Land Bank with broader authority and more tools that allow it to engage proactively as a management and development partner (Alexander 2011).

One significant legislative change in Michigan and Ohio was reform of the tax foreclosure system. If the property owner has stopped paying taxes, local government can obtain the property but it must do so in ways that are consistent with state law. In Michigan, the state created a three-year property forfeiture and foreclosure process. If a property owner has not paid taxes for two years, the property is forfeited to the county. If the property owner has not paid the third year, the property is subject to a tax foreclosure. As we said, in many municipalities the first disposal step for tax-foreclosed property is to sell the property at auction. In most cases, if it does not sell, it can be transferred to the land bank. Some land banks have the authority to identify parcels for acquisition before they are auctioned. This gives them more freedom to strategically acquire properties to bundle and later sell, or to create public amenities such as parks or green infrastructure.

Land banks are structured in different ways. Some more actively engage with and manage where they dispose of property to thwart speculative investors who have no interest in redevelopment, while others dispose of property to any willing buyer (Hackworth 2014). A comparison of Cleveland's and Detroit's land management programs showed that—even in situations with similar demand—Cleveland's approach to land management, which involved more public engagement and made more strategic decisions on how to sell properties, resulted in better redevelopment outcomes than Detroit's, which attempted to dispose of property as quickly as possible (Dewar 2006). To acquire and maintain properties for which there is no demand and to make strategic decisions about when to facilitate redevelopment, land banks need a reliable and adequate source of funding. With adequate funds they can maintain property, create green space with long-term maintenance, and acquire and bundle parcels in ways that can facilitate larger-scale redevelopment. Without ways to finance these efforts, land banks must dispose of as much property as possible, which limits long-term strategic action. Although public ownership and management can improve neighborhood outcomes when private demand is limited, how to fund land banks in financially strapped shrinking cities is unresolved.

NEIGHBORHOOD STEWARDSHIP AND DIY RESIDENT ACTION

Some of the most valuable resources in shrinking cities are its residents and local institutions such as churches and businesses. Residents are invested in their neighborhoods and, in shrinking cities, residents are on the forefront of acting as stewards for abandoned properties. One important role that residents play is that they absorb or annex abutting properties both legally and informally. By doing this, they expand the size of their lot, reducing density, and reducing the number of households to maintain the area. This practice has been called "blotting" as residents throughout the city acquire one, two, or half a dozen properties that they fence off, maintain and use (Amborst, D'Oca, and Theodore 2008). In many cases they buy the parcels from the owner or through tax sales. In other cases, residents begin to use and maintain property that has been abandoned, but some residents do not feel authorized to take over unused land even if they would be willing to maintain or use the property (Garvin et al. 2013).

To maximize the amount of property in the hands of remaining residents and reduce the amount of property they must maintain, municipalities in shrinking cities have attempted to facilitate residents maintaining more property by transferring property in public ownership to community members. Detroit's White Picket Fence program allows neighbors to buy a vacant city-owned lot next to their house for $200 (Laitner 2012). New Orleans' Lot Next Door program offers residents the opportunity to purchase any lots with which they share a lot line that are controlled by the New Orleans Redevelopment Authority (NORA n.d.). Figure 9.3 shows an example of a property acquired through the Lot Next Door program. The cost of the fencing and other improvements

Figure 9.3 **Corner lot acquired through the Lot Next Door program, New Orleans, LA. The inset shows the large yard created by the lot acquisition.**

Photograph by Casey Schreiber

was credited towards the purchase price. In neighborhoods with severe disinvestment, residents also maintain property that does not abut their property, and more ways to transfer property to residents are necessary.

As urban services decline, residents also engage in urban self-governance and begin to incorporate infrastructure maintenance and other services into their daily routine that would have been provided by local governments. Many activities occur informally, where residents take action or contribute to paying for needed services. On their own initiative, residents mow the grass, board up buildings, and organize neighborhood clean-up events. They even maintain parks and other public lands, and pay for snow plowing or other necessary maintenance. Whether in New Orleans, Buffalo, Detroit, or Cincinnati, these actions are common because governmental services declined due to reduced tax revenues. In many cases, neighbors also take defensive actions to stop scrapping (collecting scrap metal or materials for sale) and arson (Kinder 2014).

In some cases, neighborhoods set up security districts or other self-taxing districts that cover street cleaning and park maintenance as well as private security. One early district in New Orleans, the Garden District Security District, was established in 1999. The city now has more than two dozen districts. Although focusing on residential areas, they resemble the business improvement districts discussed in chapter 6 in many ways. These forms of self-provisioning are critical to maintain neighborhood quality, but they also are filled with contradictions (Kinder 2014). They are exclusionary tactics to keep decline at bay and exclude other people associated with that decline (Kefalas 2003). These tactics also reflect new conditions in the neoliberal political economy where governmental services decrease and public spaces are privatized. In the most active areas, private security in urban areas works similarly to gated communities in the suburbs.

LAND-INTENSIVE USES FOR VACANT LAND

Because shrinking cities have excess vacant parcels, and community members want to improve food access, many people from community residents to corporations have begun promoting urban farming in shrinking cities as a land-intensive industry that could stimulate local economies. Organic farms or those using integrated pest management techniques can have layered impacts including providing jobs while addressing food-related health epidemics and food deserts—the situation where a community is underserved by grocery stores selling fresh produce.

Urban farming has a long tradition in the US and occurs throughout the world. World War II was an active period for gardening in the United States. As resources and people were mobilized for the war effort, US residents experienced food rationing and shortages. The US government encouraged people to plant "victory gardens" to grow their own fruits and vegetables. The US Department of Agriculture estimates that 20 million victory gardens were planted during World War II. In 1943, three-fifths of the population grew some produce and this totaled 40% of all produce consumed (Bittman 2011). After the war, during the period of massive suburbanization, urban gardening became less common and the numbers of gardens declined substantially.

This began to change slowly in the 1970s. Beginning in New York, residents sought to control and maintain local community green space as well as grow food. A glut of vacant property during that period made it possible. Residents established groups like the Green Guerrillas in New York that transformed vacant lots into community gardens. By the 2000s, both residents and local governments in growing and shrinking cities alike have sought space for community gardens. The "global food chain" organization Urban Farming has over 62,000 partner gardens. Cities from New York to Detroit to Los Angeles have seen a surge of interest in community gardens (Urban Farming n.d.).

Researchers in Oakland, Portland, Seattle, Cleveland, Detroit, and New York, among others, have inventoried vacant land that has agriculture potential (McClintock, Cooper, and Khandeshi 2013). An analysis in Cleveland suggested that well-managed agriculture on 80% of vacant land could produce between 22% and 48% of Cleveland's demand for fresh produce, 25% of poultry and eggs and 100% of honey. The city currently supports 200 community gardens (Grewal and Grewal 2011). Community gardens also create community-based public spaces and gardening provides recreation and exercise.

In northwest Milwaukee, Growing Power shows the potential of urban farms. Fourteen greenhouses occupy the two-acre Growing Power farm. On this urban plot, Will Allen grows a quarter-million dollars' worth of produce annually which is sold through an onsite store as well as farmers' markets, and to schools and restaurants. Allen provides jobs to nearby residents, most of whom would not have had farming skills without working at Growing Power, and food to an area served by more liquor stores than grocery stores. Growing Power provides other benefits to the city. It takes in 6 million pounds of food waste in a year to compost, reducing solid waste-disposal fees and selling excess compost to farmers and residents. Growing Power also leases a 30-acre lot in the Milwaukee suburbs and has satellite offices in other cities. While Growing Power hires between 30 and 50 people, it continues to rely on volunteers and grant funding to make ends meet (Royte 2009, Growing Power n.d.).

Nevertheless, urban agriculture can be an economic development strategy. As more restaurants and residents turn to local food, the possibility for regional farms to develop a profitable niche in a world of global agricultural conglomerates increases. More than 8,000 farmers' markets operate in the US, up 76% since 2008 (USDA 2014). Community-supported agriculture is also on the rise. More individuals are returning to growing some of their own produce, and food giant Walmart recently pledged to double its sales of local produce (Bittman 2011). To develop significant and reliable yields, community gardens must be professionally managed, and both land and technical assistance are necessary to make urban agriculture a viable part of the food system. Growing Power in Milwaukee and its satellite training sites in numerous states offer just that. Facing climate change and increased energy costs, diversifying and decentralizing the food system can be one step towards developing sustainable, affordable food systems in the US and around the world as well as providing jobs in job poor neighborhoods.

CREATIVE AND TEMPORARY RESPONSES

Shrinking cities have also stimulated many creative responses to the changing conditions. In Detroit, residents attempt to make empty buildings appear occupied. They put up curtains and window shades, place furniture outside, and mow the lawn. Through their actions, otherwise unoccupied houses feel lived in (Kinder 2014). In Cincinnati, instead of creating an illusion of occupancy, Keep Cincinnati Beautiful's Arts Program paints windows and doors over boarded-up buildings. In addition to other revitalization strategies, the nonprofit organization uses this approach to improve residents' everyday surroundings, deter vandalism, and encourage investment. In the city's Over-the-Rhine neighborhood, for example, the one-time high of 50,000 residents has shrunk to 5,000, making it impossible to occupy all buildings (Keep Cincinnati Beautiful n.d.), but the paintings create a sense of life and care (Figure 9.4).

The Heidelberg Project in Detroit is a now-famous example of a multiple block art installation. Over 25 years ago, a resident and artist Tyree Guyton began making art from found objects from the streets of Detroit on the properties around his house. Using discarded everyday objects from scrap metal to toys, he created energy on vacant properties and abandoned houses across a two-block area. The project developed into a community arts organization that promotes arts education. It has also become a major tourist destination, receiving 275,000 visitors annually. Although the visitors attest to the power of the project, the City of Detroit has not always agreed and the houses that were part of the project were subject to demolition on two occasions (Heidelberg Project n.d.). Artists living elsewhere find inspiration in changing urban environments and the openness that comes with decline. In 2011 and 2012, a group of artists developed an abandoned house in New Orleans into "The Music Box, A Shantytown Sound Laboratory." The

Figure 9.4 Even though the buildings are empty, ground level façade painting prevent the buildings from looking abandoned

Keep Cincinnati Beautiful's Arts Program

Music Box was created through salvaged materials with various instruments embedded in its structure, and performances were held (Dithyrambalina n.d.).

Shrinking cities have an openness and sense of possibility that accompanies less regulation and more informality. They have street art and murals on both occupied and abandoned structures as well as temporary activities that range from pop-up shops and restaurants to flea markets and art installations. The everyday quality of the art, adaptation, and temporary uses (that we discuss further in chapter 11) sets off the gritty urbanity from entertainment-oriented destinations in the downtowns that cities are building to attract creative-class professionals. Newcomers to shrinking cities find the accompanying grit and decline inspiring because the cities feel less constraining and open to new thinking, and join longtime residents in bringing life to otherwise struggling areas. They are opening coffeehouses or restaurants, renovating houses to sell or live in, and operating businesses that sell furniture made from scavenged materials, produce from the urban farms, and handmade crafts (Dawkins 2011). It is worth noting that although shrinking cities facilitate much creative action, the abandoned property also creates stress. In 2012, nearly a quarter of Detroit's 350,000 housing units were empty. While residents appreciate the creative responses to local conditions, they report they would rather have new residents in their neighborhoods (Kinder 2014).

PLANNING TO GROW AND TO SHRINK?

Despite the extensive investment and creativity in shrinking cities, strategic action continues to be necessary. The land management and creative strategies do not address the difficulties with retrofitting utilities to smaller populations or the historic structures cities want to preserve but for which they have no immediate use. Given these challenges, many people continue to ask whether shrinking cities can attract new residents. Or conversely, are there ways to shrink to better provide a high quality of life for remaining residents? Can cities do both? Shrinking cities are no less growth-oriented than rapidly growing cities, but many have struggled to successfully rekindle their manufacturing economies or attract corporations that provide new jobs for existing residents and attract new residents. Even as cities plan to shrink, they are also planning for future possibilities, including growth.

THE GROWTH-ORIENTED BIAS

A challenge when planning for a shrinking population is the growth-oriented consensus that we discussed in chapter 5. Real estate professionals, business leaders, mayors, planners, and others involved in the urban redevelopment process make it difficult to plan effectively for shrinkage. For some, admitting to shrinking is tantamount to admitting failure. Even as a city's population declines, public and private sector actors pursue strategies for new growth rather than adapting to the reality of a smaller population. Seeing New York, Chicago, and Washington, DC regain population after a period of significant population loss makes it even more tempting to envision growth rather than plan for shrinkage.

Urban planning developed in the context of rapid urban growth and the tools that planners use often function better in growing contexts than those experiencing slow growth or population loss (Bramley and Kirk 2005). In the 2010s, the emphasis on market-based policies, consistent with a neoliberal ideological framework, makes shifting toward alternative planning tools difficult for many cities and states.

Can planning make a difference in shrinking cities? As a Sunbelt city that lost population since the 1960s, New Orleans was in an unusual situation after the city flooded following Hurricane Katrina in 2005. It engaged in extensive planning processes to chart its redevelopment path. Although residents continued to raise concerns about long-term population loss— blighted property and the loss of retail as well as poor quality public services—the planning processes did not address the reality of a smaller population. Moreover, the process made it clear that planners had few tools to address shrinkage. Their best practices—nodes of high-density development and walkable communities—would serve only some neighborhoods well. No proposals accounted for how neighborhoods would become less dense (Ehrenfeucht and Nelson 2011). A few cities including Detroit and Youngstown have begun actively to plan for their current population, a population significantly lower than the peak, but planners and city officials have not yet developed the tools necessary to implement shrinkage strategies (Schatz 2013, City of Youngstown 2010, Detroit Future City n.d.). We discuss the opportunities in Detroit's Future City plan in Box 9.1.

BOX 9.1   REVITALIZING THE RUST BELT? PLANNING A NEW DETROIT

In 2010, under Mayor Dave Bing, Detroit began to envision a new future in stark contrast to its current situation. The city had lost over 1 million residents, was $20 billion in debt, and had additional unfunded liabilities, over 70,000 blighted properties, and over 100,000 vacant parcels. What will a new Detroit look like? Should it plan to grow again or to shrink? Can it maintain its current population, given the ongoing exodus from the struggling city?

The resulting strategic framework, Detroit Future City, wavers between optimistic and realistic. It intends to stimulate the economy while addressing declining infrastructure in creative ways. The process has also had its share of controversies. The desire to plan for the future without addressing immediate concerns has resulted in a plan that many consider top down, despite an active engagement process. An idea that the city would benefit by consolidating sparsely occupied neighborhoods—although receiving positive national press—was met with resistance by residents.

By 2014, when Mike Duggan took office, the city's tone had shifted. In his *Every Neighborhood Has a Future* plan, Duggan (2014) promised to save every neighborhood and as many structures as possible while working with residents who have been working tirelessly to save their neighborhoods. In September 2013, months prior to the time Duggan took office, the Obama Administration also created a $300 million initiative to help Detroit address blight, infrastructure and public works, and public safety. The Blight Removal Task Force surveyed all of the city's 380,000 properties to build a usable database of property conditions. They found over 84,000 blighted structures and vacant lots, 98% of which were residential. Blighted property had one or more of the following conditions: public nuisance, attractive nuisance, fire hazard, tax-reverted status, under control of a land bank, vacant for five or more years, code violations, or placement on the city's demolition list.

Mayor Duggan supported saving all neighborhoods while at the same time he proposed to offer families incentives to move from sparsely occupied neighborhoods to more populated areas. While the Blight Removal Task Force's blight reduction plan supported Duggan's position that all neighborhoods must be saved, it also promoted aggressively and rapidly demolishing all blighted structures and those with indicators of future blight. The neighborhood demolition and deconstruction alone would cost $850 million. The industrial sites, although fewer than 1,000, would cost another $500 million to $1 billion because of potential contamination. The plan suggested targeting demolition efforts, and the first phase would impact 27% of the city's residents. The targeted areas were selected based on the best chance of keeping the neighborhoods stable, attracting investment, and positively impacting the largest number of residents.

What will happen to all the empty parcels and less occupied parts of the city? Demolition is one critical step, but is just the beginning. It is hard to envision a future for Detroit without growth. The plan, while attempting to adjust to a smaller population, simultaneously plans for growth. As the Blight Removal Task Force (2014) report states, the objective is to "retain and attract residents." Detroit Future City emphasizes economic growth, and Detroit Future City is working with the Detroit Economic Growth Corporation to identify and repurpose industrial properties for future economic use while also supporting entrepreneurship and small business development.

Despite the planning efforts, Detroit's future is uncharted. The plans have important tools: the Detroit Land Bank Authority and code enforcement, blight removal, and economic development initiatives. The city also has 700,000 residents working to keep the city strong. All indicators suggest it will be a smaller city with a less dense urban form. Can it also be greener and healthier, and have an economy to support its residents? Will it have a robust open space network of parks, gardens, and farms? Or are there other alternatives? While Detroit is an extreme example of population loss, it is not unique. Planning for prosperity in this context raises new challenges. Planning and revitalization techniques need to meet those challenges with new visions as well as actions that build up from the grassroots into citywide systems of change.

Source: Detroit Blight Removal Task Force (2014)

ATTRACTING NEW RESIDENTS

Cities that have lost population overall can experience growth in particular neighborhoods as suburban residents move back into central city areas. For instance, between 2000 and 2010 in St. Louis, the central corridor, an area comprising 17 neighborhoods and 50,000 residents, saw an 11% gain in population, while the 62 neighborhoods in the north city and south city areas lost 15% and 9%, respectively. The overall loss in the city was 8.3% (Ihnen 2014). Noting the trend towards urban living, shrinking cities and governors in Rust Belt states have focused on strategies that may reverse the "brain drain" and attract new mobile residents. Influenced by Richard Florida's creative class theory, cities often partner with local employers to target recent college graduates or mobile young people. The creative class theory suggests that young creative professionals move based on quality-of-life considerations and that new business follows due to the entrepreneurial activities of the creative class itself.

To further attract new residents, cities and states have developed revitalization initiatives to attract young college graduates. Launched in 2003, Michigan Governor Jennifer Granholm's Cool Cities Initiative was an explicit attempt to attract and retain a highly educated and young workforce in a state that lost residents as college-bound and educated residents moved away. The initiative initially invested $3 million in projects such as loft conversions, arts organizations, and public space redevelopment projects to create a quality of life that young people sought. The city also established a "cool city" designation that made an area eligible for grants, loans, and technical assistance (NGA n.d.). The initiative has helped support many projects that brought vitality to cities' urban cores, such as turning an old pie factory into an arts center in Saugatuck, and helping to revitalize Detroit's Eastern Market. However, the projects alone have not stemmed the general trend of population loss.

Other people-based programs have worked to attract residents who otherwise would live in disinvested communities. In the 2010s, employers involved with Detroit's Live Downtown program (which include Blue Cross Blue Shield of Michigan, Compuware, DTE Energy, Marketing Associates, Quicken Loans, and Staffing Solutions) provide mortgage and rental assistance to residents who move to targeted downtown neighborhoods (Live Downtown Detroit n.d.). The Detroit Medical Center, Henry Ford Health System, and Wayne State University have incentivized living near their facilities with their Live Midtown program (Live Midtown Detroit n.d.).

Proponents of these initiatives argue that attracting new residents, and particularly people with high levels of human capital, are critical for a city's survival. Critics argue that they direct public dollars to those who have the most opportunities in an attempt to stimulate gentrification. Benefits to long-term residents, particularly low-income residents, have been minimal and money would be better spent creating opportunities and better neighborhoods for remaining residents. Even in Detroit, with the most extensive abandonment of any US city, residents are being priced out of gentrifying neighborhoods rather than helped by targeted initiatives that attract and benefit newcomers.

In a context of overall limited demand, efforts that revitalize one neighborhood can have the effect of attracting residents from other neighborhoods, a zero-sum game (Beauregard 2013, Couch et al. 2005). We discuss this topic in more detail in Box 9.2.

BOX 9.2 IS TARGETED DEVELOPMENT EFFECTIVE IN SHRINKING CITIES? ANTICIPATING THE IMPACTS

Researchers and practitioners often propose targeted investment strategies to maximize the effects of public investments. Public entities invest in urban revitalization to help stimulate a healthy local market, and targeted investment can leverage private dollars to change local

---

circumstances. Cities often find it challenging to target resources, however. The public sector may become torn between competing demands that can pit an equitable distribution of resources against targeting proposals.

In 1999 Richmond, Virginia chose to target a majority of its CDBG and HOME funds as well as capital facilities investment, code enforcement, and other targeted action to a few select blocks in each of seven neighborhoods that showed significant signs of distress across many measures. The objective was to invest enough to stimulate private market activity that could sustain itself without public subsidy. The initiative, called Neighborhoods in Bloom, also worked with the Richmond office of the New York-based Local Initiatives Support Corporation (LISC), which provided resources and assistance to community-based organizations.

Five years later, when the effects were evaluated, the researchers found that property values in the targeted neighborhoods increased 9% faster than in the city as a whole, and areas within a mile of the targeted areas also increased faster than in the city as a whole (Accordino, Galster, and Tatian 2005, Galster, Tatian, and Accordino 2006). The Richmond in Bloom program suggests that if a critical level of investment can be made, targeted efforts can improve conditions in distressed areas. The neighborhoods had high levels of disinvestment, but it is important to note that the targeted neighborhoods in Richmond were located near downtown, and had historic housing stock and other amenities.

Would targeting work in shrinking cities? In the 1990s, Philadelphia's mayor set out to revitalize Lower North Philadelphia, one of the city's most distressed neighborhoods. It invested in building the 176-unit Poplar Nehemiah Houses and also engaged in other public–nonprofit efforts as well as a HOPE VI redevelopment of a former public housing project. The efforts demonstrated that public investment can have a great impact. They stimulated redevelopment of the area as well as gentrification, even though the city made explicit attempts to provide affordable housing in the area. As the area gentrified, housing costs in the area quickly became out of the reach of former residents who had low incomes. The efforts also did not stimulate improvements in other parts of North Philadelphia that remained affordable, however, and they continued to be distressed (Ryan 2012). Revitalizing particular neighborhoods when cities are not experiencing overall growth can lead to further property abandonment in the weakest neighborhoods (Beauregard 2013).

Targeting initiatives alone, therefore, are not a solution for shrinking cities. In shrinking cities, targeting initiatives must be accompanied by plans for untargeted areas too. If they are not, targeting could strengthen some neighborhoods while weakening others that are already struggling. Shrinking cities need innovative strategies and paradigms that address both the most and least desirable neighborhoods, or any investment will lead to disinvestment elsewhere.

RELOCATING REMAINING RESIDENTS AND CONSOLIDATING DEVELOPED AREAS

Would it be possible to concentrate all remaining residents in areas with high levels of services and amenities? The question about how to adapt a city's built form to the new realities of the smaller population has been a recurring and unresolved challenge. In 1976, Roger Starr, at the time serving as New York's Housing and Development administrator, proposed planned shrinkage in an attempt to abandon depopulated areas and consolidate remaining residents in a smaller area. The immediate opposition made it impossible to consider, and Mayor Abraham D. Beame disavowed the idea (Rybczynski and Linneman 1999, Lambert 2001). Along with 1970s New York, residents in Detroit in the 1990s and New Orleans in the 2000s have explicitly rejected shrinking the city's footprint or decommissioning neighborhoods as a solution to population loss and property abandonment (Ehrenfeucht and Nelson 2013, Hackworth 2014).

Planned shrinkage proposals raise a sticky problem in big cities. The neighborhoods with the highest levels of abandonment and that are subsequently proposed for shrinkage tend to be occupied by African American and Latino residents. Residents rightfully have argued that the plans disproportionately impact them and appear to be attempts to expel lower-income residents from the city. Popular narratives accentuate the problem. Urban planners and other interested stakeholders have used the phrase "a blank slate" to describe New Orleans after the 2005 hurricanes and in Detroit in the 2010s (Nelson, Ehrenfeucht, and Laska 2007, Dawkins 2011). While this contributes to the sense of excitement for newcomers who are predominantly white and well educated, it devalues the attachments of existing residents, rendering them invisible in the discussions about how to retrofit cities to thrive with a smaller population.

While the realities of disproportionate impacts prevented shrinkage proposals from gaining traction, other dynamics are also at play. Residents are attached to their neighborhoods and no city has yet produced a program that makes relocating to move-in ready alternative housing simple. The city of Youngstown is an example. It underwent a participatory planning process where residents and officials agreed to consolidate the population that had dropped to less than 50% of its peak. When offered relocation assistance, residents still resisted moving (Joffe-Walt 2011).

Despite past resistance, residents and public officials continue to discuss shrinking. Providing urban services such as police and fire, street lights, water and sewerage, as well as garbage collection, continue to be necessary—and difficult. Property abandonment continues to concern residents, even those who want to stay in their neighborhoods. Consolidating remaining development continues to be a rational response to high vacancy and declining municipal services—two problems that trouble residents and public officials alike. Although consolidation does not provide immediate solutions for the unoccupied land, the proposals suggest that the land no longer needed for urban uses could be allowed to return to "nature" or become regional parks or agricultural land, or de-annexed and allowed to be developed as new suburban areas (Rybczynski and Linneman 1999). However, until cities develop fair resettlement plans and residents agree to move, such proposals will likely be resisted—and as long as they are resisted, cities will allow residents to stay in all neighborhoods.

SUMMARY

Losing population across the city creates unique problems. Although the challenges resemble those facing neighborhoods in all regions, the magnitude of the problem differs. Urban shrinkage reflects regional shifts rather than absolute population growth or loss nationally. Shrinking cities can be located in regions that continue to add housing units and attract residents to the outlying cities and suburbs, but fragmented jurisdictions across the region have generally blocked regional strategies to address population loss in a major city. Municipalities that are faring better do not want to be burdened by a weak, large city.

Shrinking cities face big challenges. Population loss greatly reduces available resources, leading to reduced-quality urban services. Because shrinking cities have weak real estate markets, local governments become responsible for maintaining, securing, or demolishing abandoned parcels and structures. This becomes an increasing expense that municipalities must cover as revenues drop, but the costs of doing nothing are higher. Deteriorating neighborhood quality hurts remaining residents and can contribute to residents' decisions to move out of the city.

Vacant and abandoned property is the most visible problem and, given the overall decreased demand, envisioning cities with more land-intensive uses has become necessary. In the context of limited demand, one neighborhood can thrive even though it is surrounded by blight and specific neighborhood-scale interventions are inadequate. Visions for new citywide patterns of urban settlement are necessary, yet fraught with political challenge.

STUDY QUESTIONS

1   What are the differences between how a city shrinks and how a city grows? Why does this create challenges for shrinking cities?

2   What are the advantages of place-based initiatives in shrinking cities? What are the advantages of people-based initiatives?

3   Because shrinking cities have an excess of land, advocates suggest using this land for urban farming or other land-intensive activities, parks, and transferring land to existing properties. What are the strengths of these proposals? What are the limits?

REFERENCES

Accordino, John and Gary T Johnson. 2000. "Addressing the vacant and abandoned property problem." *Journal of Urban Affairs* 22(3): 301–315.

Accordino, John, George Galster, and Peter Tatian. 2005. *The impact of targeted public investment on neighborhood development: research based on Richmond Virginia's Neighborhoods in Bloom program*. Richmond, VA: Community Affairs Office of the Federal Reserve Bank of Richmond.

Alexander, Frank S. 2011. *Land Banks and land banking*. Flint, MI: Center for Community Progress.

Amborst, T, D D'Oca, and G Theodore. 2008. "Improve your lot!" In *Cities growing smaller*, edited by S Rugare and T Schwartz, 45–64. Cleveland, OH: Cleveland Urban Design Collaborative, College of Architecture and Environmental Design, Kent State University.

Bass, Margaret, Don Chen, Jennifer Leonard, Lisa Mueller Levy, Cheryl Little, Barbara McCann, Allie Moravec, Joe Schilling, and Kevin Snyder. 2005. *Vacant properties: the true cost to communities*. Washington, DC: National Vacant Properties Campaign.

Beauregard, Robert A. 2009. "Urban population loss in historical perspective: United States, 1820–2000." *Environment and Planning A* 41(3): 514.

Beauregard, Robert. 2013. "Strategic thinking for distressed neighborhoods." In *The city after abandonment*, edited by Margaret Dewar and June Manning Thomas. Philadelphia, PA: University of Pennsylvania Press.

Bittman, Mark. 2011. "Food: six things to feel good about." *New York Times* March 22. http://opinionator.blogs. nytimes.com/2011/03/22/food-six-things-to-feel-good-about/?_r=0

Bontje, Marco. 2005. "Facing the challenge of shrinking cities in East Germany: the case of Leipzig." *GeoJournal* 61(1): 13–21.

Bramley, G and K Kirk. 2005. "Does planning make a difference to urban form? Recent evidence from Central Scotland." *Environment and Planning A* 37(2): 355–378.

Central Terminal Restoration Corporation. n.d. "The rise and fall of rail service at the Buffalo Central Terminal." http://buffalocentralterminal.org/history

City of Youngstown. 2010. "Youngstown 2010." www.cityofyoungstownoh.com/about_youngstown/ youngstown_2010/index.aspx

Couch, Chris, Jay Karecha, Henning Nuissl, and Dieter Rink. 2005. "Decline and sprawl: an evolving type of urban development – observed in Liverpool and Leipzig 1." *European Planning Studies* 13(1): 117–136.

Dawkins, Nicole. 2011. "Do-It-yourself: the precarious work and postfeminist politics of handmaking (in) Detroit." *Utopian Studies* 22(2): 261–284.

Detroit Blight Removal Task Force. 2014. "Detroit Blight Removal Task Force Plan." http://report.timetoendblight.org

Detroit Future City. n.d. "Detroit Future City." http://detroitfuturecity.com

Dewar, Margaret. 2006. "Selling tax-reverted land: lessons from Cleveland and Detroit: New This Spring Westchester." *Journal of the American Planning Association* 72(2): 167–180. doi: 10.1080/01944360608976737

Dewar, Margaret and June Manning Thomas, eds. 2013. *The city after abandonment*. Philadelphia, PA: University of Pennsylvania Press.

Dithyrambalina. n.d. "Dithyrambalina musical architecture in New Orleans." www.dithyrambalina.com

Duggan, Mike. 2014. *Every neighborhood has a future: Mike Duggan's neighbourhood plan 2014.* www.dugganfordetroit.com/wp-content/themes/duggan/DugganNeighborhoodPlan.pdf

Ehrenfeucht, Renia and Marla Nelson. 2011. "Planning, population loss and equity in New Orleans after Hurricane Katrina." *Planning Practice & Research* 26(2): 129–146. doi: 10.1080/02697459.2011.560457

Ehrenfeucht, Renia and Marla Nelson. 2013. "Recovery in a shrinking city: challenges to "Rightsizing" post-Katrina New Orleans." In *The city after abandonment*, edited by M Dewar and J Manning Thomas, 133–150. Philadelphia, PA: University of Pennsylvania Press.

Eppig, Marianne and Lavea Brachman. 2014. *Redeveloping commercial vacant properties in legacy cities: A guidebook to linking property reuse and economic development.* Colombus, OH: Greater Ohio Policy Center.

Galster, George, Peter Tatian, and John Accordino. 2006. "Targeting investments for neighbourhood revitalization." *Journal of the American Planning Association* 72(4): 457

Garvin, E, C Branas, S Keddem, J Sellman, and C Cannuscio. 2013. "More than just an eyesore: local insights and solutions on vacant land and urban health." *Journal of Urban Health* 90(3): 412–426. doi: 10.1007/s11524-012-9782-7

Grewal, Sharanbir S and Parwinder S Grewal. 2011. "Can cities become self-reliant in food?" *Cities* June 20.

Growing Power. n.d. "Growing Power." www.growingpower.org

Hackworth, Jason. 2014. "The limits to market-based strategies for addressing land abandonment in shrinking American cities." *Progress in Planning* 90: 1–37. doi: http://dx.doi.org/10.1016/j.progress.2013.03.004

Heidelberg Project. n.d. "Overview." www.heidelberg.org

Hollander, Justin B and Jeremy Németh. 2011. "The bounds of smart decline: a foundational theory for planning shrinking cities." *Housing Policy Debate* 21(3): 349–367.

Hollander, Justin B, Karina Pallagst, Terry Schwarz, and Frank J Popper. 2009. "Planning shrinking cities." *Progress in Planning* 72(4): 223–232.

Ihnen, Alex. 2014. "Understanding population change and density in St. Louis." nextSTL.com September 17. http://nextstl.com/2014/09/pxstl

Joffe-Walt, Chana. 2011. "A shrinking city knocks down neighborhoods." *Planet Money* on *NPR* March 15. www.npr.org/sections/money/2011/03/15/134432054/a-shrinking-city-knocks-down-neighborhoods

Keenan, Paul, Stuart Lowe, and Sheila Spencer. 1999. "Housing abandonment in inner cities – the politics of low demand for housing." *Housing Studies* 14(5): 703–716. doi: 10.1080/02673039982687

Keep Cincinnati Beautiful. n.d. "Arts." http://keepcincinnatibeautiful.org/programs/urban-revitalization/arts

Kefalas, Maria. 2003. *Working-class heroes: protecting home, community, and nation in a Chicago neighborhood.* Berkeley, CA: University of California Press.

Kinder, Kimberley. 2014. "Guerrilla-style defensive architecture in Detroit: a self-provisioned security strategy in a neoliberal space of disinvestment." *International Journal of Urban and Regional Research* 38(5): 1767–1784.

Kurzius, Alexa C. 2014. "Repurposing old rail stations in the Rust Belt: what Buffalo, Detroit and Cinncinati can tell us about adaptive reuse." *Belt Magazine* September 2. http://beltmag.com/repurposing-old-rail-stations-rust-belt

Laitner, Bill. 2012. "Detroit sells city-owned vacant lots to neighbors, cleaning up neighborhoods." *Detroit Free Press* December 14.

Lambert, Bruce. 2001. "Roger Starr, New York planning official, author and editorial writer, is dead at 83." *New York Times* September 11. www.nytimes.com/2001/09/11/nyregion/roger-starr-new-york-planning-official-author-and-editorial-writer-is-dead-at-83.html

Live Downtown Detroit. n.d. "It pays to live Downtown, literally." www.detroitlivedowntown.org

Live Midtown Detroit. n.d. "It pays to live in Midtown!" www.livemidtown.org

Mallach, Alan and Lavea Brachman. 2013. *Regenerating America's legacy cities.* Cambridge, MA: Lincoln Institute of Land Policy.

Marchand, Yves and Romain Meffre. 2010. *The ruins of Detroit.* Göttingen, Germany: Steidl.

Martinez-Fernandez, Cristina, Ivonne Audirac, Sylvie Fol, and Emmanuèle Cunningham-Sabot. 2012. "Shrinking cities: urban challenges of globalization." *International Journal of Urban and Regional Research* 36(2): 213–225.

McClintock, Nathan, Jenny Cooper, and Snehee Khandeshi. 2013. "Assessing the potential contribution of vacant land to urban vegetable production and consumption in Oakland, California." *Landscape and Urban Planning* 111: 46–58. doi: http://dx.doi.org/10.1016/j.landurbplan.2012.12.009

Nelson, Marla, Renia Ehrenfeucht, and Shirley Laska. 2007. "Planning, plans, and people: professional expertise, local knowledge, and governmental action in post-Hurricane Katrina New Orleans." *Cityscape* 9(3): 23–52.

New York Times. 2014. "Detroit's fight against blight." *New York Times* June 7. www.nytimes.com/2014/06/08/ opinion/sunday/detroits-fight-against-blight.html

NGA. n.d. *Case study: Michigan's "Cool Cities" initiative*. Washington, DC: National Governors Association, Center for Best Practices.

NORA. n.d. "Land Stewardship." New Orleans Redevelopment Authority. www.noraworks.gov

Plyer, Allison and Elaine Ortiz. 2012. "Benchmarks for blight: how much blight does New Orleans have?" The Data Center. www.datacenterresearch.org/reports_analysis/benchmarks-for-blight

Richardson, Harry W and Chang Woon Nam. 2014. *Shrinking cities: a global perspective*. London: Routledge.

Royte, Elizabeth. 2009. "Street farmer." *New York Times* July 1. www.nytimes.com/2009/07/05/ magazine/05allen-t.html

Rugare, Steva and Terry Schwarz. 2008. *Cities growing smaller*. Cleveland, OH: Cleveland Urban Design Collaborative, College of Architecture and Environmental Design, Kent State University.

Ryan, Brent. 2012. *Design after decline: how America rebuilds shrinking cities*. Philadelphia, PA: University of Pennsylvania Press.

Rybczynski, Witold and Peter D Linneman. 1999. "How to save our shrinking cities." *National Affairs* 135 (Spring): 30–44.

Schatz, Laura. 2013. "Decline-oriented urban governance in Youngstown, Ohio." In *The city after abandonment*, edited by Margaret Dewar and June Manning Thomas. Philadelphia, PA: University of Pennsylvania Press.

Shrinking Cities. n.d. "Shrinking Cities." www.shrinkingcities.com

Thomas, June Manning. 2013. *Redevelopment and race: planning a finer city in postwar Detroit*. Detroit, MI: Wayne State University Press.

Urban Farming. n.d. "Urban Farming: more than a gardening organization." www.urbanfarming.org

USDA. 2014. "New data reflects the continued demand for farmers markets." US Department of Agriculture. www.usda.gov/wps/portal/usda/usdahome?contentid=2014/08/0167.xml

Virginia Tech Metropolitan Institute. 2006. *Blueprint Buffalo*. www.oneregionforward.org/plan/blueprint-buffalo

10

Cleaner and greener urban environments

INTRODUCTION

Urbanization has paradoxically led to higher living standards and, at the same time, environmental degradation that threatens the well-being of people and other species. Urban development has resulted in a host of negative environmental consequences: water pollution, "ozone action days" where it is dangerous for some people to go outside, neighborhoods that experience high levels of asthma and other diseases, and species and habitat destruction. People of color and low-income communities are disproportionately burdened by these environmental and health hazards. They also tend to have less access to parks and other environmental amenities. Environmental justice, sustainability, and climate change adaptation have become critical urban issues that require action at multiple scales. Each urban revitalization initiative is an opportunity to improve local environmental quality and move towards a systematic transformation.

In recent decades, revitalization efforts have responded directly to the environmental impacts of urban development and seek to create a new cycle of community benefits. Remediation of previously contaminated brownfield sites can provide redevelopment opportunities while reducing local exposure to toxic substances. Green infrastructure, active transportation improvements, parks, trails, and other green spaces can improve life quality as well as help cities adapt to long-term environmental sustainability. Greener environments have psychological and social benefits as well as the potential to reduce residents' household expenses and strengthen regional food systems. Environmental benefits have also become powerful marketing tools, and cities and neighborhoods tout their environmental elements to attract residents and businesses.

Environmental sustainability is a critical component for all urban revitalization, but the history of uneven development shows that it is equally important to ensure environmental improvements do not benefit only wealthy residents, or create new patterns of inequality. Instead, revitalization efforts must improve overall quality for all urban residents while transforming urban living patterns into ways of life that are sustainable across multiple dimensions. In this chapter, we give a brief history of urban environmentalism. Next, we discuss some of the specific techniques of environmental revitalization including remediating brownfields, installing green infrastructure, and improving urban ecosystems. These efforts can improve overall environmental and urban quality. We then discuss parks and recreational facilities and green industries as projects that can help revitalize neighborhoods while leading to more sustainable cities. We end with a discussion about the range of activities required for climate change adaptation.

CONTINUITY AND CHANGE IN URBAN ENVIRONMENTALISM

Improving urban environmental quality has a long tradition. As nineteenth-century industrial cities grew, reformers and municipal professionals proposed myriad improvements that ranged from water infrastructure, sewer systems, and paved streets to trash pickup and street cleaning. Efforts to humanize harsh urban environments stimulated a parks movement and led to construction of some notable US urban parks including Central Park in New York City. The visions of greener cities also influenced the creation of parkways, or tree-lined streets and suburbs intended to integrate urban and rural elements (Cranz 1982, Mohl 1988, Schultz and McShane 1978).

Urban environmental quality further improved through efforts that gained force in the 1970s. Federal laws including the Clean Air Act (1970), Clean Water Act (1972), Endangered Species Act (1973), and Resource Conservation and Recovery Act (1976) to control hazardous waste, have all positively impacted urban life and urban development. The federal government passed the Comprehensive Environmental Response, Compensation, and Liability Act (CERCLA), commonly referred to as the Superfund program, in 1980. The program created a tax on petroleum and chemical industries and a $1.6 billion trust fund to pay for the remediation of highly contaminated sites including those where the responsible party cannot be identified. To address concerns about associated costs and liability associated with redeveloping contaminated sites, in 1994 the EPA developed the Brownfields Economic Redevelopment Initiative which gave $200,000 grants to help offset the costs of brownfield assessment and remediation. In 2002, the Small Business Liability Relief and Brownfields Revitalization Act protected prospective property owners from liability when they redeveloped contaminated sites and authorized funding for cities to conduct citywide brownfield assessments.

Environmental improvements have been tremendous. The Environmental Protection Agency has helped fund thousands of brownfield assessments and cleanups. Although some business groups claim that environmental improvements are impediments to economic growth, experience with the 1970 Clean Air Act tells a different story. Emission of six common air pollutants has decreased by 72% since 1970 even as the population increased by 53% and energy consumption increased by 47%. As this occurred, the GDP increased by over 200% (USEPA 2013a). Conversely, declining environmental quality can drain the economy and increase health care costs. The air quality in Chinese cities such as Beijing met local standards on only 48% of the days in 2013, and some days are so hazardous that the US Embassy recommends people avoid all outdoor activity. In addition to impacting residents' health, the air pollution is impacting tourism and makes recruiting senior executives difficult (Wong 2014).

Researchers have nevertheless documented that environmental harms—or benefits—are not evenly distributed. In 1987, the United Church of Christ's Commission for Racial Justice's influential report *Toxic Waste and Race* showed that African American and Latino residents disproportionately lived near toxic facilities. This was just the start of extensive research and activism around environmental justice (Goldman and Fitton 1994). The research that followed repeatedly found that African American and Latino residents as well as all low-income residents disproportionately lived near toxic facilities such as chemical processing, oil refineries, or garbage dumps. They also lived in neighborhoods with fewer parks, open spaces, and pedestrian and bicycle amenities. The research continued to show how unequal social structures shape uneven exposure to environmental harms and access to environmental benefits (Bullard 2008).

Environmental justice organizations have been active in discussions about environmental regulations since that time but their successes have been mixed. Impacted communities have organized to stop toxic facilities from locating in their neighborhoods and succeeded in many cases, although not in others (Pellow 2007). Cumulatively the demands for safe and clean environments for all people have gained attention. In 1994, President Bill Clinton issued Executive Order 12898 entitled "Federal Actions to Address Environmental Justice in Minority and Low-income Populations" requiring every agency to develop an environmental justice strategy to identify and address disparate impacts on communities of color or with low incomes. In 2011, President Barack Obama issued a Memorandum of Understanding that affirmed the federal government's commitment to further environmental justice (FIWGEJ 2014). As a result, the US Department of Transportation, for example, requires all transportation projects using federal funds to avoid disparate impacts on communities of color or with low incomes, and requires that community members are involved in planning processes. In Oakland, CA, in response to community concerns that a proposed new parking structure for the Fruitvale Bay Area Rapid Transit (BART) system would create adverse impacts with no benefits to the local community, BART worked with the Unity Council, a local redevelopment corporation, to redesign the improvements to include a pedestrian plaza that would connect to a nearby commercial area (USDOT n.d.).

Ongoing concerns about pollution and environmental inequality continue as newer concerns have surfaced. Simultaneously with tackling environmental justice issues around water and air quality and hazardous waste, planners have turned to counteract the adverse health impacts from sedentary urban living habits with active transportation and physical recreation. Combatting and adapting to global climate change, however, might become an overriding concern, and reducing consumption of energy that produces greenhouse gases will be one priority. Active transportation can have multiple benefits from improving health, improving public life, and reducing fossil fuel consumption. Denser settlement patterns and transit-oriented development that reduce energy use can also have benefits such as retaining agricultural lands and wildlife habitat as well as reducing deforestation. Adapting to climate change has also become an increasing priority in many countries. Some actions therefore focus on reducing energy use and fossil fuel reliance to reduce the rate of climate change, while others seek to aid communities vulnerable to sea-level rise, storm damage, and flooding.

A PATCHWORK OF ENVIRONMENTAL IMPROVEMENTS

Much urban revitalization occurs on a project-by-project basis. As a result, we tend to see only incremental environmental improvements. Urban greening initiatives generally can be divided into efforts that explicitly aim to create green jobs, businesses, or amenities (such as a park or an alternative energy business) and those that are part of revitalization projects, but not the primary objective. In this section we discuss remediation of brownfield sites, green infrastructure, and improving urban ecosystems. Each is an important effort that can improve local environments even though greening may not be the primary objective of the initiative.

BROWNFIELD REDEVELOPMENT

Brownfields are parcels that may be contaminated with hazardous substances and toxins such as hydrochloric acid, zinc compounds, arsenic, or lead. Although many brownfields were former industrial or manufacturing sites, others include gas stations, dry cleaners, paint stores, or auto repair shops that provided everyday goods and services to a neighborhood. Brownfields may be well maintained empty parcels, or properties that do not look contaminated. Because of the expense and potential liability in brownfield reuse, many sites are left undeveloped. As a result, brownfield remediation often becomes a necessary first step in revitalization initiatives. The perception that a site is contaminated can be as much of a deterrent as actual contamination and some people call both actual and perceived contaminated sites brownfields.

Prior to reuse, the sites must be remediated or cleaned to ensure they will be safe for future users. Failing to clean sites properly can lead to ongoing water contamination or can cause adverse health impacts for people on or near the properties. However, remediation or even the potential for remediation creates a hurdle in the redevelopment process. The added costs and potential liability to future developers, particularly in areas with weak real estate markets, reduces the likelihood of private redevelopment. When approaching brownfields, planners and developers often use a "triage approach" (Frank and Sounderpandian 2001). This means they focus their limited resources on the most marketable sites that have redevelopment potential rather than reducing environmental contamination in the most affected communities.

While the triage approach may make sense from a financial standpoint, according to the USEPA more than 450,000 brownfield sites need to be remediated and redeveloped in the United States (USEPA 2012). Although the EPA has contributed funds for thousands of properties, the sheer number of sites nationally means that such an approach will exclude many sites in disadvantaged neighborhoods. Brownfields are disproportionately located in communities of color and low-income neighborhoods which are more likely to have weaker real estate markets. As a result, these communities contain brownfield parcels that can be left abandoned and undeveloped for decades.

The properties most severely contaminated with hazardous materials are eligible for listing on the Superfund National Priorities List, but these are generally not considered brownfields. In these cases, the situations are deemed so hazardous that remediation funds are available through the federal Superfund program. Although the Superfund program has made progress and remediated close to 400 sites, in January 2014 there were over 1,300 sites on the list and new sites continuously being identified (USEPA 2013b). USEPA also has a larger list of properties that were or will be evaluated for possible Superfund designation. After evaluation, those not listed as Superfund sites are given the designation "no future remediation action planned." Although likely large sites, the possibility of severe contamination has been ruled out and therefore the sites can be appealing redevelopment prospects (USEPA 2013b).

Love Canal in Niagara Falls, New York is a well-known Superfund story that illustrates the challenge of redevelopment. The site was used in 1942 by Hooker Electrochemical to dispose of 21,000 tons of hazardous chemicals. Disposal ceased and the site was covered in 1953. The land was then leased to the Niagara Falls Board of Education in 1954 and property was developed, with an elementary school. By 1978 there were 800 houses and 240 apartments nearby. From the 1950s to the 1970s, residents complained about odors and substances in their yards. Residents also experienced terrible health problems from exposure to the buried toxins. The first study, done in 1976, identified toxins on the site. After a protracted struggle, the residents were relocated and the site was listed on the Superfund National Priorities List in 1983. In 2004, $400 million and 20 years later, the site was finally removed from the Superfund list (Revkin 2013).

While still challenging, unlike Superfund properties, brownfields can be remediated and redeveloped in a reasonable period of time. Identifying a site to redevelop is the first step. If a community-based organization is instigating a revitalization effort, developing an available inventory of potential neighborhood sites can aid potential investors. If there is reason to believe sites are contaminated, CBOs can take various steps to determine the history of different sites. A title search, Sanborn maps, and aerial photographs are all sources that can help determine a site's former use and the potential for contamination. Local agencies and state brownfield authorities might have compiled lists of brownfield sites and will make this information available (APA 2010).

Once sites have been identified, the site's contamination levels must be assessed. This involves a two-part environmental site assessment. The first part, known as Phase I environmental site assessment, is conducted in accordance with the USEPA's "all appropriate inquiry" standard (ASTM E1527-05). These are the requirements for assessing the conditions of the property prior to when it was acquired. This important step can prevent a landowner from being liable for conditions that existed prior to acquisition. This assessment determines how the property was used, what substances were used, whether cleanup had occurred, where wastes were disposed of on the site, and whether there are engineering controls (such as landfill caps) or institutional controls such as deed restrictions or excavation prohibitions. It also should indicate whether nearby pollutants might have contaminated the site. Consultants often perform the assessments, and the costs range from over $1,000 to many thousands of dollars (APA 2010).

The objective of a Phase II environmental assessment is to determine as much about the contamination as possible, including the types of contaminants, the boundaries of the contamination, and the concentration levels. This will help determine what remediation is necessary. Phase II assessments are more expensive and can range from $10,000 to $100,000 or more. Phase II also includes potential cleanup plans. The cleanup can increase the property's value, and to many communities the cleanup itself and reducing the hazardous condition is as important as future development. The remediation costs will vary by the level and type of contamination and the proposed future uses (APA 2010).

Although resources are available to help remediate brownfields, property owners can be resistant to having properties labeled brownfields because it can harm development potential. Increased awareness about the health impacts of toxins from some types of business and industry has resulted in residents' opposition to locating these businesses in their neighborhoods. "Not in my back yard" (NIMBY) has been used to describe residents' responses. Political actions against toxins, and growing awareness that stopping a toxic facility will result in it being sited elsewhere, have led to the demand for NIABY—"Not in anyone's back yard." Nevertheless, many communities contain potentially hazardous faculties, and far more live with the legacy of past activity.

Successful examples of reused brownfields abound. Visitors to Victory Park in Dallas, TX or the fans filling its American Airlines Center rarely consider the site's industrial past. The 75-acre site held energy plants with associated facilities, a landfill and garbage crematorium, among other uses, that left a toxic legacy. Despite the site's prime location, the cost of remediating the site discouraged its reuse. To develop it, in the late 1990s a public–private partnership was established between the City of Dallas, the Hillwood Development Corporation, Hicks Holdings, and community residents. The Hillwood Development Corporation led the cleanup, working with the city, community, and the Texas Voluntary Cleanup Program which provides technical and legal incentives including protecting future owners and lenders from liability. The American Airlines Center was completed in 2001. It has 1,000 housing units and over 600,000 square feet of office space, and the site's development continues (Victory Park n.d, USEPA 2007).

On a smaller scale, in Portland, OR the Portland Alumnae Chapter Delta Sigma Theta Sorority converted an abandoned gas station into a green building to serve as the June Key Delta Community Center. The sorority sought the "living building designation" and not only remediated the site but also pursued green building standards that included net zero energy and water use as well as nontoxic, recycled, and locally sourced building materials. In addition, they sought to increase equity through hiring women and minority-owned firms (June Key Delta Community Center n.d.).

In other cases, former brownfields have created the opportunity for new park sites. Seattle's wildly popular Gas Works Park reminds users of the historic roots. In 1975, the city worked with landscape architect Richard Haag to turn the abandoned gas works into a lakefront park, and Haag decided to retain rather than remove the historic gasworks. Houston's Discovery Green, a 12-acre, centrally located former brownfield, cost the city $182,000 to restore as a park. It is now surrounded by a new grocery store and millions of dollars in other private investment (Harnik and Donahue 2012). Given the potential for brownfield sites and the ongoing reality that many are left undeveloped, the City of Cleveland has taken the lead and established an industrial–commercial land bank that we discuss in Box 10.1.

BOX 10.1 CITIES IN THE LEAD: THE CLEVELAND
INDUSTRIAL–COMMERCIAL LAND BANK

Once the center of the iron and steel industry, the City of Cleveland deindustrialized in the second half of the twentieth century, leaving behind a substantial amount of vacant industrial and commercial land. In 2005, Cleveland established a land bank specifically for commercial and industrial properties "as a proactive approach to reusing properties with serious real estate obstacles, such as environmental contamination and/or economic hardship" (City of Cleveland n.d.). The land bank takes the lead to assemble large parcels for future industrial or manufacturing uses, including demolishing existing structures and remediating the properties (City of Cleveland n.d.). Unlike a land bank that acquires residential and small commercial properties, an industrial land bank acquires properties with characteristics that make them suitable for future industrial uses, including adequate acreage or contiguous boundaries with other industrial sites, and locations that will help preserve existing industry such as a location by a freeway. After remediating a site, the land bank can sell it or retain it until demand for development increases. The land bank is an economic development strategy as well as a mechanism to control vacant property and remediate toxic sites. Its purpose is both to attract new industry and to reduce contaminated and blighted industrial property.

Nevertheless, the Cleveland Land Bank has faced the same challenges as other potential developers. While a site might have potential, costs of remediation are high and often uncertain. When the city acquired and remediated the Monarch Aluminum Factory, it acquired the property for only $25,000 or the back taxes owed. At its peak, the factory had 500 employees manufacturing aluminum products such as pots and pans. The factory stopped operating in the 1980s and, after some temporary uses, it stood empty, becoming an imminent threat to the neighborhood. The city's objective was to remediate the site so it would be available for new development to help revitalize the Cudell neighborhood. The costs of remediating the site were close to $3,000,000—the City paid over $2 million: $750,000 from the state and $164,000 in EPA assessment grants—when unsafe amounts of polychlorinated biphenyls (PCBs) were discovered on the property. Because this became too costly for the city, it applied for and received assistance from the US EPA Region 5 Emergency Response program, resulting in a property available for redevelopment.

These costs and risks discourage new industrial development on previously industrial sites when developers can seek uncontaminated greenfields in the suburbs. Because of the dilemma, the actions by the city's innovative land bank are necessary to attract industry back to the core. The land bank also has other benefits to residents. PCBs are dangerous to residents if they are exposed, and the city had to expend funds for fire, police, and code enforcement as the site stood empty. Nevertheless, as a shrinking city, Cleveland's population has dropped from a peak of close to a million to less than 500,000 in 2010. It has too few resources to acquire and maintain all unused vacant property and each remediation is costly. As a result, the city strategically evaluates properties for their potential market value and many properties will remain abandoned.

GREEN INFRASTRUCTURE

In cities across the United States, sewerage and storm water systems are failing to protect nearby surface- and groundwater. Energy demand and costs have increased. Utilities are struggling to meet this demand and residents are struggling to pay utility bills. These challenges have encouraged municipalities to augment traditional infrastructure with green alternatives. Green infrastructure uses natural processes to improve how urban infrastructure functions and to reduce energy use. Some examples include using parks and open spaces to collect water in retention ponds and then allowing it to infiltrate naturally through porous soils. This process cleans the water and prevents contaminated runoff from entering water bodies, and slows the rate of discharge into the drainage system, thereby reducing the number of overflow events and preventing the need for costly upgrades. Trees and other vegetation can take up water, which also slows the rate at which it enters a stormwater system. Shading also greatly reduces surface air temperatures. These forms of green infrastructure can also provide amenities while improving environmental quality. The challenge is that green infrastructure requires coordinated local and regional changes to maximize the benefits. Moreover, like brownfields, they also involve upfront costs and therefore are most likely to be installed in areas with greater market potential.

Urban areas have unique environmental dynamics. Asphalt, concrete, and roofs absorb heat during the day and release it at night, creating a phenomenon called the heat island effect. During the day, average urban temperatures for cities with more than a million people are in the range 1.8–5.4°F (1–3°C) hotter than non-urban surrounding areas. At night, when the air cools but the surfaces emit heat, urban areas can be as much as 22°F (12°C) warmer (USEPA n.d.-a). This raises cooling costs in summer, which increases energy use and pollution from energy production. It not only makes urban areas less comfortable, but also can result in heat-related hospitalizations and even deaths.

Impervious surface cover, or the streets and buildings that do not allow water to infiltrate, also greatly increase the speed and quantity of stormwater runoff. Estimates suggest that 50% of suburban land cover is impervious. In Manhattan, it is as high as 94% (Polycarpou 2010). Conventional stormwater management systems either release stormwater directly into surrounding water bodies or combine it in the sewage system. The problem is that untreated urban runoff deposits petrochemical road residues, pesticides, fertilizers, and pet waste into surrounding water bodies. Combined systems become overloaded during peak events, resulting in raw sewage flowing directly into water bodies. As little as 10% impervious surface can degrade water quality in nearby rivers, lakes, and estuaries (Stoner, Kloss, and Calarusse 2006). Increasing infiltration by slowing the rate at which stormwater goes into conventional systems can reduce contaminants from entering nearby water bodies. Infiltration cleans the water and slower runoff can reduce the likelihood and intensity of overflow events.

Green infrastructure can further help to mitigate these problems at a variety of scales (for one example, see Box 10.2). For example, downspouts may be disconnected from buildings to retain and harvest stormwater onsite. Rain gardens, bioswales, and planter boxes can also reduce stormwater runoff. Permeable paving on streets and sidewalks, and greening parking lots and streets, are shown to increase infiltration and reduce heat. Green roofs can reduce heating and cooling costs as well as reduce wear and tear on heating, ventilating, and air conditioning (HVAC) systems. Restoring urban streams and increasing available capacity for biofiltration on individual sites also contributes to an integrated green infrastructure system because it is impossible to eliminate impervious surfaces in urban areas.

BOX 10.2 PROJECTS THAT WORK: THE CASE OF THE EAST KOLKATA WETLAND

Experimentation with green infrastructure in US cities has most commonly focused on stormwater management. US cities have only slowly begun to embrace recycling and reusing water and sewage sludge. In countries around the world, however, public officials have realized that the more efficiently a city can recycle or reuse materials, the more cost-effective and reliable the infrastructure will be. The East Kolkata Wetlands, in Kolkata, India, is an innovative system that treats sewage and manages stormwater for use in fisheries and rice paddies in a functioning ecosystem (Dhar 2013).

The East Kolkata Wetlands is a 125-acre site that contains 254 fisheries fed by sewage, agriculture plots, and rice paddies as well as solid waste farms. As one effort to assist the overtaxed sewerage infrastructure, the wetlands have been designed to detain sewage that is treated by an algal photosynthesis-based system, a process as or more effective than conventional sewage treatment. The effluent from treated sewage first flows through the fishponds and then the water is used to irrigate nearby rice paddy fields. This engineered system was built within a natural wetland system to take advantage of the natural processes already at work. It is the world's largest aquaculture system, which relies on remediating and reusing wastewater and solid waste from the surrounding area. In 2002, this innovative project was recognized as a wetland of international importance by the Ramsar Convention on Wetlands (East Kolkata Wetlands Management Authority n.d.).

The wetland conservation has additional ecological benefits. The wetlands act as a carbon sink, serve as a catch basin for surface runoff, and recharge the groundwater, all while providing jobs and food. Green infrastructure is a long-term investment and the benefits increase over time. Nevertheless, it also faces pressure from ongoing development and growth to convert land to non-wetland uses. It also needs regular upkeep and investment to ensure it continues to function well (East Kolkata Wetlands n.d., East Kolkata Wetlands Management Authority n.d.).

The Milwaukee Metropolitan Sewerage District (MMSD) has become a national leader in its efforts to integrate green infrastructure into its combined system. The MMSD has implemented improvements on public land to improve surface water quality without the need to build hard infrastructure. The efforts included private property owners' onsite retention efforts and grants to help property owners install green infrastructure. The MMSD dedicated $5 million in capital matching funds during 2010–11 to support green roof grants, rain barrels, and rain gardens from property tax revenues (Garrison and Hobbs 2011). Similar programs in Chicago and Portland provide grants for property owners to install small commercial or residential green roofs and provide discounts on utility bills for properties with onsite stormwater management (Stoner, Kloss, and Calarusse 2006).

The amenities provided by green infrastructure for local residents and street users are equally important. The MMSD also worked with the City of Milwaukee to transform a three-mile stretch of the southside neighborhood into a green corridor with solar traffic signals, recycled water streams, rain gardens, trees and other vegetation, along with other amenities. Twenty residences have also added rain gardens, and one mushroom farm opened (Garrison and Hobbs 2011, Edible Milwaukee n.d.). In highly urbanized areas, cities such as Portland and Seattle have installed street planters to manage stormwater, which also provides water to often water-starved street trees (Figure 10 1). Similarly, the Village Homes community in Davis, CA created a green drainage system that prevents stormwater up to and greater than a ten-year storm event without any discharge into the municipal water system (Village Homes n.d.).

Urban street trees can be a cost-effective way to improve public aesthetics and provide savings to property owners. Canopy trees can offer shade in summer and protection from rain. Both shading and evapotranspiration can counter heat island effects, which can reduce energy use as well as make urban summer environments more comfortable. In hot climates, shaded streets are necessary to allow all residents, regardless of health or ability, to walk, cycle, or use public transit. Deciduous trees lose leaves and allow light through in colder months. Street trees can also reduce air pollution and greenhouse gases. Because of the great potential benefit and relative affordability, cities as diverse as Los Angeles and New York City in the US to Mississauga outside Toronto, Canada have developed "million tree" or other urban plant initiatives to improve urban quality through tree planting (City Plants n.d., Million Trees NYC n.d, One Million Trees Mississauga n.d.). Urban streets, however, create unfavorable conditions for trees and, as a result, many street trees have a short life span or potential benefits are never realized. Planting trees in conjunction with creating rain gardens and pervious planting strips can create better conditions for the trees while also maximizing the use of the planter strips.

Figure 10.1   Curb extensions with planters and bike lanes at SE 12th Ave and SE Clay St in Portland, OR
Photograph by City of Portland Bureau of Environmental Services

Green infrastructure nevertheless raises a new set of challenges. Creating comfortable human habitats has increasingly separated people from the local climate, and people now live in regions and have lifestyles that rely on extensive cooling and heating. In dry desert climates such as Phoenix, AZ, where daytime temperatures can reach 100°F, cooling is necessary to make the city livable for residents. Vegetation can be an effective cooling technique for outdoor areas and can help reduce energy consumption to cool buildings. Native vegetation in the Phoenix area, however, does not provide those benefits because the plants have adapted to the hot and dry conditions. As a result, using plants to decrease the heat absorbed by impervious surfaces and cooling neighborhoods also increases water use. This creates a different problem. Rapid growth in Phoenix has led to water demand that exceeds anticipated available water. Groundwater has been pumped more rapidly than it can be replenished. Phoenix also relies on water from the Colorado River. The 2012 US Department of the Interior's Colorado River Basin Water Supply and Demand Study found that the water available from the Colorado River is going to fall short of demand.

Nevertheless, the benefits from green infrastructure can be great. These include improving surface water quality, reducing the need to construct new infrastructure, and improving the functioning of existing infrastructure. Residents and businesses can benefit directly from reduced heating and cooling costs. The amenities are equally important and enhance neighborhood life, creating widespread benefits for different stakeholders. Green infrastructure strategies are easily incorporated into urban revitalization projects or can serve as components of larger neighborhood revitalization programs, but as living systems they also must reflect and adapt to local conditions.

URBAN WILDLIFE HABITATS AND BIOPHILIC CITIES

Green infrastructure can help restore urban ecosystems—the plants, animals, people, and habitats that make up an area. The patchwork of green spaces throughout a city, including street trees, yards, parks, and green roofs, provides habitat for a diverse range of species. As well as reclaiming vacant property or building new parks, landscaping projects provide an opportunity to increase urban wildlife habitats, in addition to the benefits we discussed previously. Although urban landscapes cannot be returned to conditions prior to human settlement, attention to native plants, habitat needs of regional species, and reducing the use of pesticides and herbicides can create ecosystems that support people and other living creatures.

There are many reasons to focus on urban wildlife. Researchers have found that urban residents enjoy seeing and interacting with animals in their local environments (Baur, Tynon, and Gómez 2013, Liu et al. 2013). For example, in New York City the red-tailed hawks have avid supporters, resulting in relocating pairs nesting in buildings. New York University set up its Hawk Cam in 2011 to chronicle the lives of red-tailed hawks living in Washington Square Park, which were nesting on the 12th floor ledge of NYU's Bobst Library (NYU n.d.). Not all urban species are uniformly popular, however: for example, around the country, coyotes enter urban areas and hunt domestic cats and small dogs (Draheim et al. 2013).

Proponents of *biophilic cities* emphasize that people adapted to live with nature for thousands of years and this need has not diminished. Growing evidence shows that rich natural environments support human health and well-being and provide physiological and emotional benefits. Green environments, the sounds of nature, and fresh air all create spaces that serve people's deep connections to natural systems and other living creatures (Beatley 2011). Because the majority of the world's people now live in cities, people come in contact with nature in cities—or they can be denied the opportunity if their environment does not support local ecosystems. In addition to green infrastructure, which can improve the quality of urban services and urban spaces, these are intrinsic reasons to foster robust natural habitats and well-functioning urban ecosystems.

Despite the tremendous possibilities, money, time, and political conflict continue to slow environmental initiatives. In many cases, environmental improvements are perceived to increase the costs of development or incur new costs to residents. Although benefits are substantial and lasting, the environmental benefits tend to accrue over longer periods and are not immediately visible. This creates opposition from affected businesses and resident groups alike. Moreover, in some cases pinpointing liability, or even identifying contaminated sites and the extent of contamination, can be difficult. Despite the challenges of incorporating environmentally sustainable elements into revitalization efforts, revitalization coalitions have the opportunity to pursue programs that improve environmental quality and make their cities environmental leaders.

REBUILDING GREEN CITIES AND REGIONS

As urban environmentalism has become more common, and the connections between poor quality living environments and health problems have become better known, there has been more attention to the creation of greener, healthier environments. Oftentimes, neighborhoods that experience disinvestment are also underserved by parks and recreational facilities, and unused land provides an opportunity to create new amenities. Others focus on local needs such as access to healthy food or job creation. All types of revitalization projects can envision strategies to maximize environmental benefits for a community while also accomplishing other objectives.

PARKS AND OPEN SPACE

As cities have grown larger and denser, demand for land has often resulted in neighborhoods with few parks and little green space. Revitalization efforts respond to this situation by including new parks and recreational facilities. A glut of unused parcels or an unused former industrial site can become an opportunity to add much-needed parks or recreational areas to a neighborhood. Park-like improvements can transform temporarily undeveloped land or brownfield properties into an asset. An underutilized parcel of land or unused rail corridor can inspire a coalition to pursue new recreational facilities. Improving street quality through tree planting, as we discussed above, landscaping, and active transportation infrastructure can help create high-quality public spaces available in each community.

Successful parks and open space projects abound. In some cases, they even involve repurposing old industrial infrastructure. In many cities, abandoned rail lines have become well-used bicycle and pedestrian trails that create public life and generate business activity along their borders. One of the most high-profile examples of this type of revitalization activity is New York's High Line Park. The 1.45-mile linear park in park-poor West Manhattan was constructed out of an abandoned elevated freight rail line that had slowly been taken over by grasses and other vegetation (see chapter 6). In addition to providing a park for passive recreation, it used 161 native species and functions like a green roof, retaining 80% of the stormwater onsite to water the plant beds, and cooling the local environment while providing habitat for insects and birds (Friends of the High Line n.d.).

The former El Toro Marine Base has been slowly transformed into the Orange County Great Park in Irvine, CA. Once fully complete, the park will cover 1,300 acres, nearly twice the size of New York's Central Park. Like the High Line, the park replaces defunct urban infrastructure with a host of new recreational amenities and a three-mile wildlife migration corridor that will link the Cleveland National Forest to its north with the Crystal Cove State Park on the other side of the park. The park also includes large groundwater recharge areas, capture areas for urban runoff, and green street designs such as porous paths and parkways and dark sky lighting.

Although parks offer desired green spaces, they also require ongoing operating expenses that rarely can be recouped by onsite uses. Large-scale park provision and restoration efforts usually involve at least one public entity. In some cases, local parks departments will take over operation, or will work collaboratively with community groups to provide additional parks. Just as often, however, residents must develop an operating plan and raise funds for continued maintenance.

In the case of the $7.7 million restoration of Nine Mile Run in Pittsburgh, PA, green infrastructure was used to restore 2.2 miles of an urban stream while extending and enhancing Frick Park, a natural and recreation area, by 100 acres (Figure 10.2). Instigated by the US Army Corps of Engineers, the project included construction of wetlands for both wildlife habitat and storm infiltration, and used Pittsburgh's porous soils to help capture surface water. The wetlands served as an infiltration area that reduced pollutants from a nearby 238-acre mountain of slag, a legacy of Pittsburg's industrial past, from entering the stream and degrading water quality. The restoration efforts removed previous engineering solutions and returned the stream to more natural functioning. It includes both conservation areas and increased recreational trails. The 2013 State of the Watershed report showed significant improvements to the water quality although it remains degraded. The project also contributes to realizing Pittsburgh's larger Riverfront Development Plan to increase the use of the riverfront while simultaneously restoring it (Nine Mile Run Watershed Association n.d.).

FOOD SYSTEMS

Food system planning intersects with many revitalization possibilities in many places. People in downtowns and the suburbs, and in growing and shrinking cities alike, are rethinking regional food systems. In chapter 9, we discussed urban farming as a possible land-intensive use for excess land in shrinking cities. This is part of a larger movement to rebuild the regional food system in order to increase food security for residents while making our food systems less environmentally damaging. Urban food production at different scales—from household food production and community gardens to urban farms or larger-scale agriculture—has been seen as a solution to various urban needs and is an important part of regional food system planning. Local food production can use vacant land, provide community green space, provide jobs in areas with excess vacant land, and provide food in food deserts or areas underserved by

Figure 10.2   Nine Mile Run in Pittsburgh, PA
Photograph by John Moyer

conventional grocery stores. A regional food system can reduce energy use by reducing food transport distances. To better facilitate local food production, cities are revising their zoning codes and local regulations to make it easier to grow and sell food in neighborhoods as well as to keep poultry or other small livestock. However, many cities still have prohibitions on selling food that is grown in city limits or community gardens, and on outbuildings on property without a primary structure.

Urban farms also can be a job-training opportunity. In Duluth, MN, Community Action Duluth's green jobs initiative Seeds of Success trains people in high-poverty areas to grow vegetables to sell in the Lincoln Park Farmers' Market. By teaching people growing skills, Seeds of Success also strives to improve households' food security by helping households grow a portion of their own food (Seeds of Success n.d.). The Lincoln Park Farmers' Market also allows vendors to rent on a week-to-week basis to sell produce from home or other local gardens, making the connections between sellers of all sizes and buyers. Another program, Growing Farms, is a northwest Minnesota farm incubator that began in 2011 to grow and support sustainable farms specifically on the urban edge. They work to link small farmers to markets and transition into sustainable farming practices (Growing Farms n.d.).

Although farming and agriculture can be considered environmentally friendly activities, pesticide and herbicide use in community and household gardens can be high and can lead to local contamination. They are also toxic to beneficial insects and wildlife as well as animals and children. There are many less toxic farming alternatives (USEPA n.d.-b), however, and growing awareness about the problems with pesticide use has increased demand for organic agriculture. Spending on organic food reached $28 billion in 2012, up 11% from 2011. Organic sellers also receive higher returns on their products (Greene 2013).

Even with all its benefits, urban farming is only part of the solution to a greener and more just food system. Some observers warn against putting too much emphasis on local food production without paying attention to the quality of jobs being created and larger structural changes necessary for a more sustainable system (Born and Purcell 2006). In many cases it is difficult to earn good wages and most urban farming job-training efforts rely on grants and additional funding sources. And while shrinking cities might have land to grow and export food, few other urban areas have sufficient land.

In addition, most urban residents are and will continue to be dependent on grocery stores to be well supplied with a reliable source of affordable food. A growing body of research shows that residents in low-income areas are often underserved by grocery stores and other sources of quality, affordable fresh food. Often facing time and transportation constraints as well, residents in food deserts have a harder time acquiring healthy food. In addition, the little fresh food and produce available is often more expensive in underserved areas. To counter this trend, the City of New Orleans partnered with the Hope Credit Union's Hope Enterprise Corporation and the Food Trust to finance supermarkets and other fresh food stores in low-income, underserved communities. The Fresh Food Retailer Initiative created access to a mix of interest-bearing and forgivable loans through the use of the city's CDBG funds that are matched by Hope Credit Union (Hope Credit Union n.d.). The objective is to increase the number of grocery stores. The Food Trust was also a partner in the Pennsylvania Fresh Food Financing Initiative that brought 88 new stores to underserved areas. Neighborhood organizations, urban planners, and municipal governments all have become involved in food system planning to achieve many objectives. Increasing food security by strengthening growing, selling, and buying options for residents can be an important dimension of revitalization efforts.

GREEN INDUSTRIES

We often approach economic growth and environmental quality as incompatible tradeoffs that are difficult if not impossible to plan for simultaneously. In his influential article entitled "Green cities, growing cities, just cities?" Scott Campbell (1996) argued that planners work at the tensions generated among these objectives, and that they often consider environmental compromise a necessary cost of prosperity. With the growing attention to green industries and the "green-collar" workforce, this is slowly changing. Environmental sustainability requires jobs and industries where individuals and regions prosper while also restoring and retaining ecological integrity.

Green industries are those that prevent harm to the environment, rely on sustainable materials, or involve environmentally sustainable design and production methods. They capture a diversity of industries and jobs, ranging from environmental consultants and lawyers specializing in environmental law to trade workers such as electricians who install solar panels, or people who install building weatherization. They include LEED-certified (Leadership in Energy & Environmental Design) architects and green building construction, alternative or renewable energy production such as solar or wind energy, as well as the manufacturing of energy-efficient products or transportation equipment such as wind turbines or electric vehicles. They also include agriculture or manufacturing processes that use renewable and recycled resources, city recycling programs, and even retailing in environmental products.

In revitalization efforts, cities and neighborhoods often seek to attract industries that will both provide good jobs and enhance environmental quality. Green industries provide a potential revitalization strategy because they are often locally rooted and involve adaptation of ordinary jobs and existing skills. The weatherization of historic buildings to make them more energy efficient, or outfitting them with alternative energy systems, are good examples. Retooling established or locally dependent industries such as transportation, construction, and manufacturing around alternative energy is another approach. Portland, OR has targeted its "clean tech cluster" consisting of wind and clean energy and green building in its local economic development plan. The city's economic development agency has developed strategies to support and link existing metal industries in the manufacturing of wind turbine parts. It has also developed green building standards to encourage the application of sustainable building techniques and the use of green building products and design firms.

Brownfield sites also can be converted to green industrial uses. Industrial uses require less extensive cleanup than housing, retail, or parks. The Philadelphia Navy Yard encompassed a 1,200-acre shipyard for almost two centuries. Seven acres of the site contained an incinerator and landfill contaminated with heavy metals. This brownfield was converted into another industrial site and now supports the largest photovoltaic installation in Philadelphia, generating enough power to serve 300 homes. The Philadelphia Navy Yard is one of many efforts to turn brownfields into alternative energy, and the USEPA has established "RE-powering America's Land" to encourage the development of renewable energy on current and formally contaminated lands (Levitan 2011).

CLIMATE CHANGE MITIGATION, CLIMATE CHANGE ADAPTATION

Climate change and its impacts has become a critical topic that cuts across all types of planning and urban development. Climate refers to the typical patterns of temperatures, precipitation, humidity, wind, and seasonal change. Climate change refers to measured changes to observed patterns over sustained time periods. One aspect of climate change, global warming, has received the most attention. Rising levels of carbon dioxide and other heat-trapping gases are warming the planet more rapidly than at any time in the past. In the US, average temperatures have risen across the contiguous 48 states since 1901 and the rate of change has climbed in the past 30 years. Seven of the warmest years on record have occurred since 1998. Global warming is accompanied by more extreme weather events and changing snow and rainfall patterns. The warmer climate will intensify extreme events such as hurricanes or typhoons, droughts, storms, rainfall, and floods. Some regions are predicted to become dryer and others wetter. Melting ice and snow cause sea-level rise that is impacting coastal areas. The impacts from this change are wide-ranging, and in different regions pressing concerns range from food and water access to vulnerability of coastal communities (USEPA 2014).

No singular type of action can reduce greenhouse gas emissions and adapt developed areas to anticipated changing weather patterns or land mass change. Climate interventions occur at different scales. The Kyoto Protocol was an international agreement linked to the United Nations Framework Convention on Climate Change among countries to reduce greenhouse gas emissions. The meeting establishing the protocol occurred in 1997 and the period for reductions began in 2008. In 2013, the United Nations Framework Convention on Climate Change in Warsaw established a timeline for the next agreement. The negotiations focused on two challenges: mitigation, or actions to reduce climate change primarily by reducing greenhouse emissions; and adaptation, or the steps taken to adjust to climate change. These include the impacts of sea-level rise, extreme weather such as the 2013 typhoon that hit the Philippines, and ecological and agricultural changes from the rising temperatures. At the national level, much focus has been on industrial emissions. California Assembly Bill 32 (AB32, or the Global Warming Solutions Act), discussed in chapter 8, is a good example of a state effort to curb greenhouse gas emissions.

Most revitalization efforts make local change, and there are ways that new development can reduce energy consumption. Revitalization efforts that create higher densities with opportunities to reduce auto trips can also reduce energy consumption and greenhouse gas production. Many advocates have proposed transit-oriented development (TOD) patterns to reduce fossil fuel consumption. Transit-oriented development incentivizes new construction and development around transit hubs. Although seen primarily as a way of curbing low-density development and car use, TODs can create walkable, transit-focused mixed-use development in areas in need of revitalization. This can bring new residents to an area while connecting existing residents to job centers (Cervero 2004). Urban planning strategies focused on green building techniques can also reduce energy use. Planners can mitigate activities generating greenhouse gases by planting more trees, restoring natural areas, and conserving remaining forests.

More immediate questions focus on how urban areas adapt to the effects of climate change. In some areas, necessary adaptations might be severe, such as reducing or forbidding development in areas vulnerable to sea-level rise or flooding. Other actions concentrate on reducing water use, diversifying food sources (as food prices rise or become scarce in some areas), planting native vegetation, and installing green infrastructure. Coastal Florida and coastal Louisiana have acknowledged that reducing coastal erosion and restoring the coast are critical components of buffering communities from storm surge. This is central to the Miami-Dade Climate Action Plan and the Louisiana Coastal Master Plan. No single action can adapt urban lifestyles to the varied impacts of climate change, but revitalization initiatives must anticipate the future effects.

SUMMARY

Revitalization projects provide wonderful opportunities to help shift urban areas to greener futures. Across cities, most environment improvement will take place through project-by-project changes, and revitalization efforts provide opportunities to add green elements, whether by a small step like adding street trees to a block, or a large effort such as turning a former industrial site into a source for green energy jobs. Revitalization initiatives add needed parks and recreational areas and help provide food security as well as opportunities for local residents. Much urban greening comes about in incremental and at times invisible ways. Brownfield remediation reduces toxins in local environments, and street trees can improve the experience of being on the street. While at times appearing incremental, collectively these efforts greatly increase sustainability as well as saving residents and businesses money through reduced energy costs. Green revitalization might increase immediate development costs, but facing global climate change and rising costs of energy and infrastructure provisions, the costs of doing nothing are greater.

STUDY QUESTIONS

1   In a nearby neighborhood, identify potential brownfields or areas that at some point contained hazardous sites. Find the Sanborn map for the area and go back through time to identify historic uses such as gas stations, cleaners, auto mechanics, manufacturing uses, or other uses that used potentially hazardous materials. Choose one site that you identified and find out as much as you can about whether the site is a brownfield.

2   Does your city have a green infrastructure plan? Do specific districts in your city? If so, examine a plan to determine what it includes. If not, investigate what green infrastructure regulations or programs your city or region has adopted. Are there specific areas where the city can do more?

3   Choose a new development project in the region. Determine if any green infrastructure elements were incorporated into the site design. Do you feel that the project adequately incorporated green infrastructure? Explain your response.

REFERENCES

APA. 2010. *Reuse: creating community-based brownfield redevelopment strategies*. Chicago, IL: American Planning Association.

Baur, Joshua WR, Joanne F Tynon, and Edwin Gómez. 2013. "Attitudes about urban nature parks: a case study of users and nonusers in Portland, Oregon." *Landscape and Urban Planning* 117: 100–111. doi: http://dx.doi.org/10.1016/j.landurbplan.2013.04.015

Beatley, Timothy. 2011. *Biophilic cities: integrating nature into urban design and planning*. Washington, DC: Island Press.

Born, Branden and Mark Purcell. 2006. "Avoiding the local trap scale and food systems in planning research." *Journal of Planning Education and Research* 26(2): 195–207. doi: 10.1177/0739456X06291389

Bullard, Robert D. 2008. *Dumping in Dixie: race, class, and environmental quality*, 3rd Ed. New York: Westview Press.

Campbell, Scott. 1996. "Green cities, growing cities, just cities?: urban planning and the contradictions of sustainable development." *Journal of the American Planning Association* 62(3): 296–312.

Cervero, Robert. 2004. Transit-oriented development in the United States: experiences, challenges, and prospects. In *Transit Cooperative Research Program*. Washington, DC: Transportation Research Board, National Research Council.

City of Cleveland. n.d. "Cleveland's industrial commercial land bank: a national model for brownfield redevelopment." www.rethinkcleveland.org/About-Us/Our-Initiatives/Industrial-Commercial-Land-Bank.aspx

City Plants. n.d. www.cityplants.org

Cranz, Galen. 1982. *The politics of park design. A history of urban parks in America*. Cambridge, MA: MIT Press.

Dhar, Sujoy. 2013. "Wetlands transform a city's sewage through a bit of solar alchemy." New City http://nextcity.org/daily/entry/wetlands-turn-a-citys-sewage-into-the-water-that-irrigates-its-crops

Draheim, Megan M, Katheryn W Patterson, Larry L Rockwood, Gregory A Guagnano, and E Christien M Parsons. 2013. "Attitudes of college undergraduates towards coyotes (*Canis latrans*) in an urban landscape: management and public outreach implications." *Animals* 3(1): 1–18.

East Kolkata Wetlands. n.d. "East Kolkata Wetlands: A 360° view." http://kolkatawetlands.org/about

East Kolkata Wetlands Management Authority. n.d. "East Kolkata Wetlands." www.ckwma.com

Edible Milwaukee. n.d. http://ediblemilwaukee.com/tag/milwaukees-green-corridor

FIWGEJ. 2014. "Environmental Justice." Federal Interagency Working Group on Environmental Justice. www.epa.gov/compliance/ej/interagency/index.html

Frank, Nancy and Jayavel Sounderpandian. 2001. "Brownfields triage: using decision analysis techniques to identify brownfields for investment." Unpublished paper, University of Wisconsin.

Friends of the High Line. n.d. "Friends of the High Line." www.thehighline.org/about/friends-of-the-high-line

Garrison, Noah and Karen Hobbs. 2011. *Rooftops to rivers II: green strategies for controlling stormwater and combined sewer overflows*. New York: Natural Resources Defense Council.

Goldman, Benjamin A and Laura Fitton. 1994. *Toxics [sic] wastes and race revisited. an update of the 1987 report on the racial and socioeconomic characteristics of communities with hazardous waste sites*. Washington, DC: Center for Policy Alternatives.

Greene, Catherine. 2013. "Growth patterns in the U.S. organic industry." US Department of Agriculture, Economic Research Service. www.ers.usda.gov/amber-waves/2013-october/growth-patterns-in-the-us-organic-industry.aspx#.VCjCDfmSxBl

Growing Farms. n.d. "Welcome to Growing Farms." www.growingfarms.org/ourfarms.html

Harnik, Peter and Ryan Donahue. 2012. *Turning brownfields into parks*. New York: Trust for Public Land

Hope Credit Union. n.d. "New Orleans Fresh Food Retailer Initiative." http://hopecu.org/business/loans/nola-fresh

June Key Delta Community Center. n.d. "June Key Delta Community Center." www.key-delta-living-building.com

Levitan, Dave. 2011. "Brown to green: a new use for blighted industrial sites." *Environment* 360 (23 June).

Liu, Xiangping, Laura Taylor, Timothy Hamilton, and Peter Grigelis. 2013. "Amenity values of proximity to national wildlife refuges: an analysis of urban residential property values." *Ecological Economics* 94: 37–43.

Million Trees NYC. n.d. "Home." www.milliontreesnyc.org/html/home/home.shtml

Mohl, Raymond A. 1988. *The making of urban America*. Wilmington, DE: Scholarly Resources.

Nine Mile Run Watershed Association. n.d. "History." http://ninemilerun.org/about/history

NYU. n.d. "NYU Hawk Cam." www.livestream.com/nyu_hawkcam

One Million Trees Mississauga. n.d. "Help us plant trees." www.onemilliontrees.ca

Pellow, David N. 2007. *Resisting global toxics: transnational movements for environmental justice*. Cambridge, MA: MIT Press.

Polycarpou, Lakis. 2010. "No more pavement! The problem of impervious surfaces." *State of the planet*, July 13. http://blogs.ei.columbia.edu/2010/07/13/no-more-pavement-the-problem-of-impervious-surfaces

Revkin, Andrew C. 2013. "Love Canal and its mixed legacy." *New York Times*, Retro Report November 25. www.nytimes.com/2013/11/25/booming/love-canal-and-its-mixed-legacy.html?pagewanted=all

Schultz, Stanley K and Clay McShane. 1978. "To engineer the metropolis: sewers, sanitation, and city planning in late-nineteenth-century America." *Journal of American History* 65: 389–411.

Stoner, Nancy, Christopher Kloss, and Crystal Calarusse. 2006. *Rooftops to rivers: green strategies for controlling stormwater and combined sewer overflows*. New York: Natural Resources Defense Council.

US Department of the Interior. 2012. "Colorado River Basin water supply and demand study." US Department of the Interior, Bureau of Reclamation. www.usbr.gov/lc/region/programs/crbstudy.html

USDOT. n.d. "Fruitvale BART Transit-Oriented Development Project, Oakland, California." US Department of Transportation. www.fhwa.dot.gov/environment/environmental_justice/case_studies/#l6

USEPA. 2007. *Victory Park – Dallas, Texas. Brownfields to greenfields*. US Environmental Protection Agency. www.epa.gov/earth1r6/6sf/pdffiles/victoryparksuccess2007.pdf

USEPA. 2012. "Brownfields and land revitalization." US Environmental Protection Agency. www.epa.gov/brownfields/basic_info.htm

USEPA. 2013a. "Air pollution and the Clean Air Act." US Environmental Protection Agency. www.epa.gov/air/caa

USEPA. 2013b. "National Priorities List." US Environmental Protection Agency. www.epa.gov/superfund/sites/npl/index.htm

USEPA. 2014. *Climate change indicators in the United States*, 3d Ed. EPA 430-R-14-004. US Environmental Protection Agency. www.epa.gov/climatechange/indicators

USEPA. n.d.-a. "Heat island effect." US Environmental Protection Agency. www.epa.gov/heatisland/index.htm

USEPA. n.d.-b. "Organic agriculture." US Environmental Protection Agency. www.ers.usda.gov/topics/natural-resources-environment/organic-agriculture.aspx#.U2LINXb69v8

Victory Park. n.d. "About VP." http://victorypark.com/about-vp

Village Homes. n.d. "About Village Homes." www.villagehomesdavis.org/public/about

Wong, Edward. 2014. "Most Chinese cities fail minimum air quality standards, study says." *New York Times* March 27. www.nytimes.com/2014/03/28/world/asia/most-chinese-cities-fail-pollution-standard-china-says.html?_r=0

11

Rebuilding people-oriented places

INTRODUCTION

What makes some places especially exciting and memorable? Why do particular public spaces beckon and feel comfortable while others discourage people from spending time or stopping? A big part of the answer relates to a dynamic urban life where people are engaged in diverse activities. Simply put, people want to be around other people. Residents and visitors alike enjoy having reasons to come out. A temporary art installation or mural can make a familiar street seem vibrant and interesting, and give passersby a reason to pause. Regular farmers' markets, public art, and special events such as festivals and outdoor music series can be regular meeting places that attract people to a city's public spaces. They can also bring new customers to local businesses. Public life is created by people, and the most engaging urban life comes about because people are out and about, using the city and its public spaces. Creating the conditions where daily public life flourishes, therefore, is an important component of urban revitalization.

In this chapter we discuss ways in which planners and community stakeholders can create people-oriented spaces that support public interactions and enjoyment. Revitalization initiatives can take advantage of many opportunities. Some activities do not need specific investment but simply the removal of outdated regulations and barriers to allow different individuals to take action—whether to start a business or play music. In the next subsection, we highlight the intense interest in public spaces in the twenty-first century. We then discuss the creation of people-oriented urban spaces. Following this, we examine the role of temporary uses and mobile activities, such as street vending, in urban revitalization. In some cases, activities that appear incompatible lead to tensions about who should have the right to use a space. Even in these situations, attempting to control what occurs, or even actively mediating among conflicting uses, can be too much. In this section, we also discuss the pros and cons of regulatory action. Finally, we outline arts-led revitalization and creative placemaking techniques that demonstrate how communities are rebuilding around arts and culture in order to retain their distinct qualities and increase livability while generating economic revitalization.

EVERYDAY PUBLIC LIFE IN THE TWENTY-FIRST CENTURY

For centuries, urban areas have been centers of the economy as well as social and cultural life. The people and activities in cities drew other people attracted by the excitement and opportunities available there. People experience the city through its public spaces. Cities themselves have been characterized as works of art and places of leisure (Olsen 1986), where daily life constitutes, in Jane Jacobs' (1961) words, a sidewalk ballet. The city is a designed environment composed of architecture and neighborhoods, parks and monuments, and it encompasses thousands of daily activities that create spontaneous interactions and the joy of being in public. Jane Jacobs' *The death and life of great American cities* (1961) and William H. Whyte's *The social life of small urban spaces* (1980) convinced generations of urban planners and public officials that more, rather than fewer, people in public make cities great.

In the twenty-first century, the appreciation of everyday life has grown. Many books show how urban public spaces come alive and how people use and adapt them. *Everyday urbanism* (Chase, Crawford, and Kaliski 2008) encourages us to pay close attention to how people adapt the city to suit their needs, whether they sell goods out of their house or set up temporary restaurants or garage sales on vacant lots. *Loose space* (Franck and Stevens 2007) and *Insurgent public space* (Hou 2010) show how people use space in unplanned ways, and highlight the enjoyment they gain from ephemeral activities, for example public performances or exercise and dance in public spaces. *The temporary city* (Bishop and Williams 2012) considers impermanent uses such as pop-up bars and restaurants or art installations that help satisfy residents' desire for change while alleviating problems associated with inevitable change that leads to the temporary underuse of urban sites. *The informal American city* (Mukhija and Loukaitou-Sideris 2014) demonstrates how people develop ways to live and work outside of formal job and housing markets through work such as street vending. Informal economic activity is untaxed and unmonitored by the government. Each of these books captures something that planning is only beginning to recognize: the do-it-yourself (DIY) activities of individuals and groups that are changing the social and economic dynamics of urban places. DIY actions shift the focus from consumer-citizens who purchase what they need to those who produce their own goods, grow food, or in other ways do it themselves.

At the same time, despite the popularity of everyday public life and the growing appreciation of informal urbanism, controversies regularly accompany public activities. Municipalities rarely plan proactively for street use. Instead, they regulate street activity and operate under a framework of traffic logic, where street professionals see the street through the objective of facilitating unimpeded travel (Blomley 2011, Ehrenfeucht 2012). In the 1980s, when many cities experienced population loss and disinvestment, unwanted street activities symbolized urban decline, and cities renewed their efforts to eliminate street users. Rather than addressing conditions leading to poverty and resulting in visible homelessness, urban residents and municipal governments alike developed "quality-of-life" campaigns that attempted to eliminate activities such as panhandling, and designed public spaces to deter undesirable activity such as sleeping in public (Vitale 2008). The "broken windows" hypothesis suggested that minor infractions lead to violent crime, resulting in cities across the US attempting to regulate street life as a solution to bigger issues (Kelling and Coles 1996). Although the broken windows hypothesis has been largely discredited, and no evidence links everyday "disorder" with crime, cities including New York and Seattle have attacked street-level activities when developing revitalization initiatives in the 1990s and 2000s (Gibson 2004, Vitale 2008).

Too many regulations and efforts to "clean up" the streets can lead to overly scripted environments. Many fear that planning for controlled public spaces will create urban life that feels commercialized and homogenized, without a sense of authenticity. Responding to Jane Jacobs' *Death and life of great American cities*, Sharon Zukin titled her 2010 book *Naked city: The death and life of authentic urban places*. *Naked city* articulated a common desire for authentic urban places but also a common concern: as people seek authentic urban places, they fear that redevelopment processes are creating themed Disney World-like environments. As cities appear more similar to one another, they lose some of the intangible qualities captured in the word "authenticity." This reflects the fact that urban experiences are increasingly provided by corporations, leading to greater homogeneity, but also reflecting standards and products that urban residents expect and want. In response, residents seek out local artisans and events unique to their city or region (Heying 2010). It is difficult to define authenticity or to develop a formula to make places feel more authentic. Many urbanites nevertheless "know it when they see it." Allowing informal urbanism to thrive can better create situations where the experiences develop from and by a city's people.

PUBLIC SPACES

How do we plan for public life? To maximize the possibilities for urban activities, cities and neighborhoods need high-quality public spaces. These can include parks and plazas, but it is important to remember that streets and sidewalks are the most extensive public spaces and those that almost everyone comes in contact with daily. Jan Gehl (2011) has observed that it is possible to influence how people use public spaces through better public space design that responds to regional, societal, and climatic factors that influence public space use.

People are more comfortable in spaces that are human scale. Vast open spaces, blank walls, long distances between buildings, and poor visual access can make a space feel less welcoming. Gehl (2011) found that people interacted more in spaces without walls, and where they walked shorter distances and traffic moved at slower speeds. Public spaces also need to be designed with the local climate in mind. In hot climates, shade or other protection from the sun will be necessary (although remember the tradeoffs in dry climates, discussed in chapter 10), as it is in wet climates. Protection from the weather by trees and awnings can make both streets and parks more comfortable. Design features that allow interactions and visual contact between abutting buildings and nearby public spaces can facilitate interaction as well as visual interest. Density also matters, and more rather than fewer activities attract people. Heavy traffic, however, can also deter use along a given street.

It can be useful to think about public spaces as *infrastructure*, *everyday spaces* and *destinations* (Ehrenfeucht and Loukaitou-Sideris 2010). In many revitalization processes, opportunities arise for street and other public space improvements. Paying attention to all three dimensions will maximize how usable and comfortable public spaces become. Basic *infrastructure* improvements can make streets and parks easier to use for travelers and other street users. Pedestrian and bike infrastructure, including bicycle facility markings, sidewalk improvements, curb cuts, street trees, wayfinding signage, trash receptacles, and good connections through parks or other public spaces can increase the convenience for nonmotorized travelers. Benches, street trees, occasional shelters, and restrooms are critical for residents who must use the facilities regardless of weather and for long periods where they might become tired. The facilities should be designed for users of all abilities and ages.

To plan for *everyday spaces*, public space planners need to observe what types of people and activities use the area. They also need to be aware of the area demographics to determine if some residents are not using public space, and consider ways to engage them. Are there many seniors who spend time in public daily? In that case, are there shaded spaces to sit, relax, play games, or talk? Are there street vendors? In that case, are there sites where vendors can sell without disrupting traffic flow? Are there places for customers to consume their food quickly? Close observation and discussions with different stakeholders can identify the best ways to plan for street activities. Small improvements to the street environment, such as strategically planting trees or installing benches, can make a big improvement to the overall quality of the public realm. The Chicago Loop, for example, has many of these features. The downtown streets are filled with planters that have low flat ledges for sitting, bike racks, and trash cans. The numerous corporate plazas provide tables and seating areas that are accessible to the street (Figure 11.1).

Figure 11.1   Planters can also provide seating and a space where food truck customers can wait: Chicago, IL
Photograph by Renia Ehrenfeucht

The street furniture also creates a buffer along the curb where food truck customers can congregate on the side of pedestrian traffic.

Some districts are planned for intensive use. Restaurant rows and entertainment districts are examples of *destinations* where enough activities occur that people come to shop or eat, but also to be around others. In those cases, more street improvements, signage, and wider sidewalks can help an area accommodate diverse people and activities. However, neighborhood parks can serve as destinations on a smaller-scale by providing a space that brings residents together for different outdoor activities or just to relax.

The objective of public space planning is to facilitate activity rather than dictate particular uses. Karen Franck and Quentin Stevens (2007) use the term "loose space" to describe the degree to which people are able to engage in spontaneous activities. Spaces range in their degree of looseness. A prison or courthouse would be examples of "tight" public institutions with little possibility for unintended and unsanctioned activities. Parks and streets also have a range of looseness. One park might be open to various activities throughout the day and into the night, providing lights and keeping the space available to users, whereas others might have strict hours, locked gates, and regulations about what can occur and where.

Observations across different situations demonstrate that people use public spaces in unintended ways and that the best planning allows people to use and adapt public spaces in ways they want (Franck and Stevens 2007, Hou 2010, Loukaitou-Sideris and Ehrenfeucht 2009). Games, exercise classes, sitting, yoga, sports, public performance, and street vending can occur in different situations. Even if a public weekly exercise class forces pedestrians to walk around, pedestrians are adaptable and enjoy seeing what others are doing. In the suburbs and central cities alike, residents and visitors are experiencing a resurgent interest in urban life. The experience of urban life is best defined by being in public.

MOBILE AND EPHEMERAL ACTIVITIES

Public spaces come alive with people. Parks and streets as well as public spaces such as libraries or school grounds can also be temporarily transformed into special event venues. Mobile and temporary events—whether they are businesses and services or public festivals—are popular because there are mutual benefits for people and businesses. Businesses such as food trucks and pop-up bars can add life to a neighborhood while providing opportunities for entrepreneurs with little start-up funding.

PUBLIC EVENTS AND FESTIVALS

Public, private, and community sector groups regularly plan festivals and parades that attract people to public spaces. Almost every city has special events, whether a Fourth of July celebration along the waterfront or a St. Patrick's Day parade. In the United States, New Orleans exemplifies this public festival culture. The city is renowned for Mardi Gras, which encompasses two to three weeks of celebrations that include staged events and planned parades alongside a myriad of informal activities throughout the city. New Orleans also holds many annual music festivals, New Year's celebrations including the Vietnamese Tet Festival and fireworks along the Mississippi River waterfront, community-based parades called second lines on Sundays, and a variety of neighborhood festivals (Figure 11.2). The city is home to numerous art markets, farmers' markets, and many others as well.

Public events have different purposes. Some are intended to be community-oriented and the cost is paid from public funds, sponsors, or the sponsoring organization. Others are large-scale events with significant economic impact and have direct revenues from entrance fees. New Orleans' ten-day-long Jazz and Heritage Festival has an estimated economic impact of $300 million and it attracts 400,000 attendees (Lopez 2013). It began more than 40 years prior as a much smaller, community-oriented event organized to celebrate regional heritage.

Weekly events can create community rituals. Farmers' markets are increasingly popular nationwide. In 2014, the US Department of Agriculture (USDA 2014) counted 8,268 farmers markets in the United States, a 76% increase since 2008. Industry estimates found that consumers spent $7 billion in direct sales from farmers to consumers in 2012, up from only $1 billion in 2005. In large cities, businesses located near farmers' markets also increase sales (USDA 2013). Farmers' markets provide a range of benefits. They link regional food producers directly with consumers, supporting regional food markets and bringing fresh food into neighborhoods. Consumers appreciate knowing where their food comes from and the social dimensions of visiting the farmers' markets. Arts markets and health fairs are other common ways that communities bring residents and businesses together.

Unless there are direct entrance fees, the revenues from public events are diffuse while the direct costs must be paid upfront. Public events can have significant expenses for permits, security, temporary seating, temporary toilets, as well as hiring musicians or play equipment. The economic impact from small events may not exceed the costs even if the social benefits do. The direct expenses require community-based organizations, nonprofit organizations and public agencies to seek private partnerships and corporate sponsors.

Public events have different planning requirements. Most are collaborative, where organizers coordinate with other groups that will be involved. Planning might involve multiple city agencies responsible for permitting and various regulations. Planners also must consider the details of

Figure 11.2   Paddlers on the water at the Midcity Bayou Boogaloo neighborhood festival, New Orleans, LA
Photograph by Renia Ehrenfeucht

various activities, anticipated attendance, hours and location, and advertising revenues and costs. Public events can be satisfying community-oriented activities. Even though some become economic drivers, this often does not occur in the short term.

FOOD TRUCKS AND STREET VENDORS

In the 2000s, food trucks and other street vendors gained wide popularity in US cities (Shouse 2011, Wessel 2012), and some, including Los Angeles, Portland, New York, Miami, New Orleans, and Chicago, began to revise their ordinances to allow street vending. Street vending and street food are commonplace throughout much of the world. In US cities, late nineteenth- and early twentieth-century streets were filled with children selling newspapers and flowers, and fresh vegetable and fish vendors traveling from neighborhood to neighborhood. Vendors bought and sold a range of products from rags to knives and other household goods. On downtown streets, prepared foods such as peanuts, tamales, and fried potatoes were available. However, around the turn of the previous century, street vending came under attack from shop owners and business leaders. Since that time, US cities have regulated or restricted street vending (Bluestone 1991, Loukaitou-Sideris and Ehrenfeucht 2009, Morales 2000).

Nonetheless, street vending continued to be a viable option for entrepreneurs throughout the twentieth century because start-up costs are low and there are few barriers to entry. Street food is also popular with the public because it provides convenient and affordable food to consumers (Morales and Kettles 2009, Shouse 2011). However, because of municipal restrictions, street vendors often operate in the informal economy. Working without permits, vendors are subject to fines and having their goods confiscated. The regulations create a risky environment. This is particularly the case for immigrant entrepreneurs who may not have authorization to work in the US and if arrested could face legal troubles and deportation.

With the new demand for street vending, virtually every major city in the US is now home to numerous food truck operators. In many cities, such as in Los Angeles, Chicago, and New Orleans, food trucks have moved beyond the neighborhoods they traditionally serve and into downtown areas, university districts, and entertainment centers (Lovett 2013). New operators have also entered the market. This new crop of food vendors are less willing to work in the informal economy. They have put pressure on city councils, demanding legitimate permits to legalize their business and the opportunity to operate without harassment. In all cases, from traditional taco trucks to new vendors, food trucks are popular because they provide food that people want and are convenient and often affordable. They also provide unique gathering spaces where people come together around their shared interest in food.

The current pressures to limit food trucks often come from business owners who have a storefront where food trucks operate. Three main issues arise in the discussion about whether and how food trucks should be permitted to operate: competition, traffic, and trash. Small restaurants consider food trucks unfair competition because their operating costs tend to be lower, allowing them to offer less expensive food. Business owners also complain that the street vendors are visible when they are on the street and block access to storefront businesses and that customers leave litter on the street. The City of Chicago has attempted to address pedestrian traffic and garbage concerns by adopting an ordinance that designates specific sites for street vending and prohibits food trucks from operating within 200 feet of a food establishment unless they are within a designated site. In some instances the 200-foot limit creates severe limitations on vendors due to the density of food establishments in the city's downtown neighborhoods. Since the 1980s, the City of Portland has made space for food trucks. When compared with many cities, the restrictions are few and it allows permitted vendors in many situations. Some food truck clusters or pods have over 30 vendors. Most nevertheless set up on private property rather than on public streets (Newman and Burnett 2012, Lovett 2013). Even though these efforts seek to "balance" competing uses, they still privilege shops and restaurants over mobile vendors.

Street vending is only one public activity subject to regulations. Street art, panhandling, skateboarding, sleeping, and busking are also common ways in which people adapt urban spaces. Nevertheless, many cities have limits on where and when street performers can perform. They have noise ordinances that make live music difficult, and frequently prohibit activities such as skateboarding and sleeping in public spaces. Regulations, more often than not, are for the benefit of property owners and businesses at the expense of public space users who are vulnerable to being moved along, fined, and at times arrested.

Are regulations the best way to plan for public space use? The evidence suggests that ordinances are popular political solutions for city councils, but that they do not resolve conflicts in a way that is satisfactory for all stakeholders. As a result, public space regulations may be ineffective planning tools. William H Whyte's extensive observations of public spaces offer a better direction for public space planning. His work demonstrated that when public spaces are well used, the potential for disruptive behavior is diminished (Whyte 1980, 1988). In other words, in many cases the problem is too few public space users, rather than the wrong public space users. Active use is the best solution against one group dominating a space. People often sort themselves out when using public spaces (Low, Taplin, and Scheld 2005) and often get along relatively well (Anderson 2011). Planning for more activities rather than specific activities might solve more problems than regulations. It is also important to keep in mind that all the public space users are also stakeholders whose interests must be incorporated into local initiatives. Municipal governments cannot actively plan for all activities that might occur. Luckily, this is unnecessary because people readily adapt public spaces, find places to sit, and use the city in varied ways. Simply allowing people to use and adapt public spaces for anticipated and unanticipated activities is an often overlooked pathway. In Box 11.1, we discuss the public space strategy of self-organization.

BOX 11.1 REDUCING REGULATIONS: SELF-ORGANIZATION AS A PUBLIC SPACE STRATEGY

Public spaces are governed by complex and often outdated regulations that can make it hard for residents to make creative use of public spaces. Changes to the city's regulatory environment that respond to contemporary street life can encourage rather than impede public actions. By revisiting street vending and parade permit requirements, produce selling restrictions, and noise ordinances, cities have been improving the conditions that facilitate public life. In some situations, the best way to plan for public life is to remove regulatory barriers that limit what activities can occur.

This can work because pedestrians and other public space users have a tremendous capacity for self-organization. Even without stringent regulations, public spaces usually function well. Even though we take it for granted, walking is a complex activity that involves constant interactions with other people and objects. In all but the most sparsely populated environments, in order to get around, pedestrians constantly make adjustments such as steering around others to retain their fluid motion and avoid collisions. For decades, observers have noted how pedestrians successfully negotiate rapidly moving and varied pedestrian environments without formal regulations (Goffman 1971, Helbing et al. 2001, Whyte 1988). The eminent sociologist Erving Goffman (1971: 11) referred to "traffic codes" as a set of "ground rules that provide normative bases of public order" on sidewalks and streets. Even though these activities are self-organizing, they're also predictable. Pedestrians trust in others' behavior while engaging in voluntary cooperation and automatically adjust routing behaviors when facing obstacles (Goffman 1971: 17). Urban planner and researcher William Whyte (1988) confirmed these findings in a multiyear research project involving countless hours of unobtrusive direct sidewalk observation, time-lapse photography, and filming to analyze pedestrians and public space use in multiple countries. As a result of the project, Whyte (1988: 56) called the pedestrian a "marvelously complex and efficient" transportation unit. Because of this, when people are walking they can maneuver environments that have other activities occurring. In fact, pedestrians in crowded environments negotiate obstructions better and more efficiently.

Self-organization has been observed in other circumstances, including among public market vendors and customers, and food truck operators and customers (Ehrenfeucht 2013, Morales 2010). On Chicago streets, even the food truck customer lines adapted to pedestrian flow. The lines bent so that a travel lane was created, or the line divided so pedestrians could walk through (Ehrenfeucht 2013). As a result, even when the food trucks attracted large crowds, they had minimal impacts on the pedestrian movement.

In addition, pedestrians and other public space users enjoy what Whyte (1988) has called the pleasures of downtown streets. In other words, walkers enjoy their surroundings and both unexpected and regular interactions in the pedestrian environment (Goffman 1971, Lofland 1998, Stevens 2007, Whyte 1988). Whyte's observations demonstrated that "what attracts people most, in sum, is other people" and too few people hurt public environments more than too many people" (1988: 10). Whyte went so far as to argue that what cities "need is pedestrian congestion" (1988: 7). Most sidewalk congestion, however, is self-congestion. People hold conversations, stop for protracted goodbyes, and pause to observe something in the midst of other people without moving out of the way. Pedestrian traffic moves around them and these sources of congestion resolve themselves. Because we have observed public space self-organization in different circumstances, because congestion resolves itself and people enjoy being around others, a new strategy to approaching street life might be simply to let public activities be. We can allow people as pedestrians, vendors, performers, customers, and audiences to adjust to others with whom they share public spaces.

THE TEMPORARY CITY

All cities have leftover spaces, blank walls, and underused properties. This is particularly true in areas with weak real estate markets where revitalization efforts can occur over a long period of time. In these instances, temporary uses such as markets or art installations can inject life into an otherwise dead and underutilized space. They can attract people and give new energy to a place while reducing the opportunities for vandalism and other potentially destructive activities. Unexpected public space infrastructure such as street furniture or spontaneous performances can also energize the street or make a privatized plaza more welcoming to other people. Flea and art markets, grassy parks, and beer gardens can all be appealing uses for unused spaces.

The past decades have seen a great increase in pop-up retail and restaurants. Pop-up retail can range from a temporary clothing store that allows a new designer a short-term, affordable lease in a storefront that otherwise would be empty, to mobile clothing stores that accompany special events. Pop-up retail and restaurants can be an affordable way to try out a business idea and allow the operator to gain a reputation prior to opening a regular establishment. In New Orleans, for example, Pizza Delicious began as a pop-up in 2011, using a shared kitchen that offered space for bakers and other independent operators to sell pizza-to-go on Sunday nights. After their popularity led to a kickstarter campaign to raise funds for a restaurant, Pizza Delicious opened as a pizzeria at the end of 2012.

In the US and UK, artists and property owners are initiating temporary land-use activities instead of leaving their building vacant (Bishop and Williams 2012). This is advantageous for other businesses in an area because empty storefronts can be a drain on retail corridors. For example, since 2001 the Phantom Galleries program has enabled artists to fill empty storefronts in and around downtown San Jose. The program, which was spearheaded by a pair of local artists and went on to receive city support, provides exhibition opportunities for local artists while benefiting the city and property owners by making streets and public spaces more interesting and attractive. The artistic transformations playfully advertise the spaces to potential occupants, too. The program has since spread to cities around the country. The blog Pop-Up City (popupcity.net) documents dozens of other whimsical and functional urban interventions that use space in innovative ways, such as gardening on train station rooftops for waiting passengers in Japan (De Boer 2014).

Other forms of temporary use are politically motivated and seek to "test alternative futures, offer new avenues for dialogue and education, and/or question urban development policy" (Ashley 2015: 6). Many projects are initiated by grassroots communities and arts groups. For instance, Rebar, a collective of artists, designers, and activists based in San Francisco, initiated Park(ing) Day in 2005 as a statement against the lack of parks and green space in downtown San Francisco. The now annual event has spread to hundreds of cities around the world in which people transform parking spaces into often eclectic public spaces (Bishop and Williams 2012, Merker 2010). Residents in Dallas, TX spawned the Better Block movement, which seeks to inspire alternative futures for auto-oriented streets through temporary interventions such as pedestrian and bike infrastructure, landscaping, and other amenities. Other examples involve public–private partnerships. In San Francisco, public and private entities work together on "Sunday Streets" to close stretches of streets to traffic and turn them into a play area for an afternoon where children and adults alike can come and enjoy the event.

Pop-ups, parklets, and other temporary land uses satisfy consumers' curiosity for newness and offer a powerful means of demonstrating new opportunities for urban spaces. Limited availability increases attention and keeps the business feeling exciting. Because pop-up bars, shops, and restaurants create a sense of excitement, something different and possibly fleeting, temporary activities have emerged in weak and strong markets alike, with and without formal sanction.

EMBRACING THE ARTS

Arts and cultural activities are another important source of urban revitalization that can activate public spaces and bring vitality to urban areas. Museums exhibiting grand artwork, unsanctioned murals, restaurants displaying the mastery of chefs and pastry makers, theaters and music clubs supporting actors and musicians, along with dozens of other endeavors reflect a city's arts and culture. As we discussed in chapter 6, cities have attempted to rebuild their downtown areas by developing major museums and performing arts centers. The arts also operate in cities in less visible ways. Artists and musicians are attracted to cities where they find the freedom, space, and audiences that enable them to pursue their interests. In turn, they build a creative milieu that at once supports their career development and stimulates public life.

An artistic presence—whether a community art center, artist coop, music venue, or public art installation—has the potential to contribute intangible benefits to communities by bringing out a locality's distinct qualities, animating underutilized spaces, and providing opportunities for different people to get involved in their neighborhood. They also have the potential to serve as an economic development driver. While cities around the world have embraced and supported the arts for their economic potential, it is important to be realistic about what they can accomplish in terms of economic development and not to overlook their role in creating interesting places to live and work. The arts are special in that they can help to create dynamic places, enhancing an area's livability in more ways than traditional economic development targets such as business services or high-tech industries. However, different types of arts activities play different revitalization roles. Some excel at attracting tourists and others are better suited to generating community engagement. Some arts activities are more likely to engender gentrification and displacement while others spur revitalization. Below, we discuss some of these variations in more detail.

ARTS IN ECONOMIC DEVELOPMENT

A number of studies describe and document the different roles of the arts in urban revitalization (Grodach 2011, Grodach and Loukaitou-Sideris 2007, Grodach, Foster, and Murdoch 2014, Markusen and Gadwa 2010, Markusen and Schrock 2009, Stern and Seifert 2010). First, the arts may provide a neighborhood anchor or amenity that attracts visitors to an area. Tourism, from the regional to the international scale, can become a significant way for an area to capitalize on its distinctiveness and local cultural traditions. It also creates opportunities for residents to spend money on local art, music, and cultural venues rather than spending money in another city or neighborhood. The arts are unique in this regard too because they can be an important form of import substitution. In other words, rather than frequenting national chain stores that sell imported goods made elsewhere, residents have the opportunity to purchase products made and designed locally.

By generating local spending and retaining local dollars, the arts also exhibit a strong multiplier effect, or the level of spending that ripples through the local economy. This is because artists are often closely tied to each other and support their local arts scene. Local art spaces incubate artistic talent and stimulate creative environments. This can generate more jobs, local spending, and creative output that will create more economic activity. This is in keeping with the trends of local businesses overall. In 2009, Civic Economics compared the local circulation of dollars from local businesses along Magazine Street in New Orleans with a chain store and found that locally owned businesses generated twice the local economic activity that chain stores generate (Urban Conservancy and Civic Economics 2009). In turn, as local artists and craftspersons grow in success, their work can become important exports for a city. The designer crafts and microbrews of Portland, OR are good examples (Heying 2010).

As we elaborate below, the arts also enhance an area's distinctiveness by helping to create more interesting streets and public spaces and affirming local traditions. In this regard, artists of all sorts rehab old buildings and start new businesses that define a city, as Heying (2010) shows in Portland. The arts may also lay the foundation for neighborhood revitalization because they provide a venue for community engagement and outreach. Arts organizations can also offer community education programs, work with marginalized groups such as the homeless and mentally ill, and spearhead public art projects (Grodach 2011). Their engagement provides an important impetus to build social ties in a community. In fact, neighborhoods with strong arts activity tend also to have stronger levels of community participation and other signs of economic revitalization (Grodach, Foster, and Murdoch 2014, Stern and Seifert 2010). Given this potential, arts-based community revitalization programs have spread around the US (Box 11.2).

BOX 11.2 CAN THE ARTS SPUR COMMUNITY REVITALIZATION? THE NATIONAL ENDOWMENT FOR THE ARTS AND CREATIVE PLACEMAKING

Creative placemaking is a relatively new movement in urban cultural policy that focuses on place-based, arts-led revitalization (Markusen and Gadwa 2010, Nicodemus 2013). The concept emerged from the National Endowment for the Arts (NEA) Our Town program, which was initiated to fund arts-based community revitalization efforts by artists and arts organizations. Creative placemaking is now championed by ArtPlace, a collaboration of 13 foundations and six banks, as well as various state and local governments and foundations. Between 2011 and 2013, NEA and ArtPlace alone made 232 grants totaling $41.6 million (Nicodemus 2013). Creative placemaking programs have funded a wide range of projects focused on arts engagement, entrepreneurship, cultural planning, urban design, and public art with the intention that "the arts will play an explicit and intentional role in helping to shape communities' social, physical and economic futures" (ArtPlace 2014). Proponents argue that creative placemaking "animates public and private spaces, rejuvenates structures and streetscapes, improves local business viability and public safety and brings diverse people together to celebrate, inspire and be inspired" (Markusen and Gadwa 2010: 3).

These programs mark a significant expansion of the NEA's previous policy of targeting grants to artists and arts organizations "for art's sake" because the NEA now also promotes the arts as a means of facilitating positive community change. Conceptually, creative placemaking is aligned with the wider placemaking literature associated with planning giants such as Jane Jacobs, William Whyte, and Jan Gehl, who promoted people-centered design and attention to the role of urban form in shaping how we use places. They also encouraged community-driven change over the top-down decision-making that defined modernist planning in their day.

Creative placemaking is also rooted in the many studies that document the role of the arts in catalyzing neighborhood change—in terms of both community engagement and providing a symbolic dimension to place that can generate economic growth and higher property values. The program seeks to encourage artists and arts organizations, in partnership with other community stakeholders, to engage in improving their communities by finding new uses for vacant buildings and underused public spaces, thereby engendering social engagement and generating economic revitalization. In fact, a key element of the Our Town program is an emphasis on cross-sector collaboration among local residents, governmental agencies, nonprofit organizations, and private sector interests.

In their study of dozens of creative placemaking techniques, Markusen and Gadwa (2010: 15) found that the major challenges are similar to other revitalization efforts. They include "creating partnerships, overcoming skepticism on the part of the communities and public leaders, assembling adequate financing, clearing regulatory hurdles, ensuring maintenance and sustainability, avoiding displacement and gentrification, documenting progress and developing performance metrics."

There is also an inherent paradox in creative placemaking policies and similar efforts to harness the power of the arts in neighborhood change. While the arts are associated with bringing many neighborhood improvements, they also may help stimulate gentrification and the displacement of lower-income residents.

In fact, recent studies problematize creative placemaking and arts-based neighborhood planning assumptions regarding the arts' potential to balance economic growth with social and community development. A national study of the arts' neighborhood-level impacts finds that while the presence of museums and fine arts companies tends to encourage slow and steady revitalization, commercial arts industries such as film and design services tend to be found in rapidly gentrifying neighborhoods (Grodach, Foster, and Murdoch 2014). Another recent study of arts organizations in New York City finds that most seek out and, by extension, support creative class destinations rather than play a direct role in communities with significant disadvantaged or immigrant populations (Murdoch, Grodach, and Foster 2015). Those arts organizations that are positioned for community revitalization tend to be the small and volunteer-driven.

The NEA has funded a wide range of projects. While some look like strong community revitalization initiatives, others appear to be geared more toward tourism or creative class consumption (Nicodemus 2013). These studies show that creative placemaking programs need to focus more closely on the type of arts activities and neighborhoods that they invest in and seriously consider the potential catalytic effects of the arts once a creative placemaking program is under way.

PUBLIC ART

Vitality in the built environment comes from many sources—architecture, landscaping, and urban design. A sometimes overlooked source of urban vitality is the sanctioned and unsanctioned art in public places. For decades, public art has been an important way to add interest and depth to public space and a means of building community pride. Public art programs support commemorative monuments, murals, abstract sculpture, environmental art, or even interactive hands-on work such as Anish Kapoor's *Cloud Gate* in Chicago. Popularly known as "the Bean," the highly polished stainless steel sculpture attracts hordes of tourists to come watch its reflections of public life in Chicago's Millennium Park.

In the United States, both state and municipal governments encouraged the production of public art by establishing "percent for art" programs. These programs require that a certain percentage of capital costs for public development projects, usually 1%, be dedicated to the commission and installation of art that is available for public enjoyment. The Philadelphia Redevelopment Agency established the first percent for art program in 1959. Today, hundreds of cities and at least 25 states operate their own program of some kind (NASAA 2013).

Public percent for art programs can provide important revitalization benefits by enlivening an area and giving people a reason to visit. For example, thousands of visitors each year go to Millennium Park to experience the Bean, or Providence, RI's WaterFire, a temporary exhibition and festival of bonfire "sculptures" on the city's downtown rivers. These visitors in turn spend money while in the area at local shops, restaurants, and hotels. This consumer spending can potentially attract new businesses as well. However, public art can also become a source of exclusion. Percent for art programs may fund installations that only improve the appeal of privatized public spaces and therefore primarily benefit developers and shoppers rather than all people living and working in an area (Sharp, Pollock, and Paddison 2005, Whybrow 2011).

Sometimes major redevelopment projects generate sizable public art budgets that can support more than a set of individual public art pieces. For example, in Los Angeles, the Community Redevelopment Agency funded the construction of the $23 million Museum of Contemporary Art (MOCA) in the Bunker Hill redevelopment area through its public art fund. For this project, the Agency revised its public art fund to extract 1.5% of development costs from developers for construction of the museum building. The redevelopment agency's objective was to help create a flagship museum that would provide an anchor in the redevelopment project that could catalyze development in the surrounding area. This creative use of public arts funds helped establish a world-renowned museum, but it failed to catalyze development as planned. This is because funding a museum building was not enough. Among other factors, the Agency did not take into account urban design and siting issues that would enable the museum to fulfill its role as development catalyst (Grodach 2008).

STREET ART

Informal and unsanctioned art can have an important role in urban revitalization as well. Writing and painting on walls has a long history, stemming from early human societies when pictographic stories were inscribed on walls. In the 1920s, during a time of social unrest after the Mexican Revolution, three great Mexican muralists—Diego Rivera, José Clemente Orozco, and David Alfaro Siqueiros—created great depictions of everyday lives and struggles. Many Chicano/a neighborhoods continue the tradition of adorning building walls with elaborate murals.

In the 1960s and 1970s, graffiti became a notable phenomenon particularly in deindustrializing US cities. Men and some women in Philadelphia and New York—notably African American and Latino/a youth—transformed walls and trains into striking graphical displays (Cooper and Chalfant 1984). Graffiti captured people's imagination and it was rapidly adopted worldwide (Iveson 2010, Reiss 2007). Street art became a global art phenomenon as writers and artists in different cities and countries developed distinct styles and expressive forms (Austin 2010, Reiss 2007). In 1979, Dan Witz hit the streets of lower Manhattan with his hummingbirds (Witz 2010) and in 1981, Blek le Rat took his art to the streets of Paris (Prou and Adz 2008). The interest in street art has only grown in the past decades. In the 2000s, both the Tate Museum in London and the Museum of Contemporary Art in Los Angeles exhibited street art (Figure 11.3). Artist Skyler Fein's "Youth Manifesto" immortalized a lighted version of the tag HARSH that was shown in the New Orleans Museum of Contemporary Art in 2009. In 2011, New Orleans Museum of Modern Art commissioned a piece by street artist Swoon.

Street art and graffiti differ from murals and other public art in an important way. Street art is unsanctioned art, and artists and graffiti writers do not ask permission to write on the walls. They might work on public or private property, but in either case they do not ask permission. In some cases, property owners immediately remove the art. In others, the property owners value the piece and intentionally retain it, at times even covering the piece with Plexiglas (Ehrenfeucht 2014). Art can be a source of tension in urban areas, where some consider unauthorized art property defacement and others consider it an asset.

In Philadelphia, city officials have worked with nonprofit organizations to channel energy focused on street art into mural projects. Beginning in 1984, as part of the Philadelphia Anti-Graffiti Network, then Mayor Wilson Goode approached muralist Jane Golden to engage with graffiti writers. Ten years later, the Philadelphia Anti-Graffiti Network was reorganized into Philadelphia Mural Arts Program in order to direct street artists' and graffiti artists' talent and interest towards sanctioned spaces. Now a public–private partnership, the Mural Arts Program has generated over 3,600 murals in the city (Mural Arts Program n.d.). For some artists and writers, this is an acceptable compromise. Others, however, are quick to note that part of street art form is to place art without asking permission and choose spots that appeal to the writers' objectives, audience, and the features of the built environment. Although mural programs can be

Figure 11.3   Street art enlivens the street, Los Angeles, CA
Photograph by Renia Ehrenfeucht

effective vehicles to introduce more art into public space, they are unlikely to eliminate all unauthorized street art.

SUMMARY

To revitalize urban areas, we must create the conditions where everyday life can flourish in public. Residents and visitors seek ways to enjoy their neighborhoods, connect with other people, and experience what is familiar and new. Activities that give cities vitality can be informal or planned, instigated by grassroots groups or local governments and chambers of commerce. An important part of planning for vibrant cities is to encourage conditions in which urban life can unfold. It requires stepping back and providing a framework where individuals can make decisions about how and when to act, and allowing for more than prescribed activities. Small public space improvements, sanctioned and unsanctioned art, addressing outdated regulations, and enabling temporary establishments and special events give people reasons to come out in public.

STUDY QUESTIONS

1   Examine the municipal code for the city you live in. Identify all the ways that public space activities are regulated. Are the regulations open or exclusive? Are they necessary? Can you imagine different ways to regulate similar activities?

2   Search the local paper's online archives for a controversial public space activity such as panhandling, street art, street vending, noise ordinances, day labor sites, or loitering. What sparked the controversy? What are the different positions?

3   Many cities have adopted "Complete Streets Ordinances" to rebuild streets in ways that support all travel modes. Find a Complete Streets Ordinance. Determine what kinds of activities are planned for and what might be omitted. In your perspective, will the streets be complete?

4   Identify an arts- or culture-oriented district in your city or nearby that was developed as a means of urban revitalization. Describe its qualities. Is it porous or closed? What types of arts activities are located there? Who lives and works there? What physical and social factors contribute to its qualities? Do you believe the district has been successful as a revitalization strategy?

REFERENCES

Anderson, Elijah. 2011. *The cosmopolitan canopy: race and civility in everyday life*. New York: WW Norton & Company.

ArtPlace. 2014. "ArtPlace America invests an additional $14.7 million in the field of creative placemaking." www.artplaceamerica.org/download/file/fid/7502

Ashley, Amanda Johnson. 2015. "The micropolitics of performance: pop-up art as urban communication." Unpublished paper.

Austin, Joe. 2010. "More to see than a canvas in a white cube: for an art in the streets." *City* 14(1/2): 33–47.

Bishop, Peter and Lesley Williams. 2012. *The temporary city*. London: Routledge.

Blomley, Nicholas K. 2011. *Rights of passage: sidewalks and the regulation of public flow*. Abingdon, UK: Routledge.

Bluestone, Daniel M. 1991. "'The pushcart evil': peddlers, merchants, and New York City's streets, 1890–1940." *Journal of Urban History* 18(1): 68–92.

Chase, John, Margaret Crawford, and John Kaliski. 2008. *Everyday urbanism*. New York: Monacelli Press.

Cooper, Martha and Henry Chalfant. 1984. *Subway art*. New York: Holt, Rinehart and Winston.

De Boer, Joop. 2014. "Japanese commuters grow veggies on train station rooftops." http://popupcity.net

Ehrenfeucht, Renia. 2012. "Precursors to planning regulating the streets of Los Angeles, California, c 1880–1920." *Journal of Planning History* 11(2): 107–123.

Ehrenfeucht, Renia. 2013. "Food trucks in Chicago." Unpublished paper.

Ehrenfeucht, Renia. 2014. "Art, public spaces, and private property along the streets of New Orleans." *Urban Geography* 35(7): 965–979. doi: 10.1080/02723638.2014.945260

Ehrenfeucht, Renia and Anastasia Loukaitou-Sideris. 2010. "Planning urban sidewalks: infrastructure, daily life and destinations." *Journal of Urban Design* 15(4): 459–471.

Franck, Karen and Quentin Stevens. 2007. *Loose space: possibility and diversity in urban life*. London: Routledge.

Gehl, Jan. 2011. *Life between buildings: using public space*. Washington, DC: Island Press.

Gibson, Timothy A. 2004. *Securing the spectacular city: the politics of revitalization and homelessness in downtown Seattle*. Lanham, MD: Lexington Books.

Goffman, E. 1971. *Relations in public: microstudies of the public order*. New York: Basic Books.

Grodach, Carl. 2008. "Museums as urban catalysts: the role of urban design in flagship cultural development." *Journal of Urban Design* 13(2): 195–212.

Grodach, Carl. 2011. "Art spaces in community and economic development: connections to neighborhoods, artists, and the cultural economy." *Journal of Planning Education and Research* 31(1): 74–85.

Grodach, Carl and Anastasia Loukaitou-Sideris. 2007. "Cultural development strategies and urban revitalization: a survey of US cities." *International Journal of Cultural Policy* 13(4): 349–370.

Grodach, Carl, Nicole Foster, and James Murdoch. 2014. "Gentrification and the artistic dividend: the role of the arts in neighborhood change." *Journal of the American Planning Association* 80(1): 21–35.

Helbing, Dirk, Peter Molnar, Illes J Farkas, and Kai Bolay. 2001. "Self-organizing pedestrian movement." *Environment and Planning B* 28(3): 361–384.

Heying, Charles H. 2010. *Brew to bikes: Portland's artisan economy*. Portland, OR: Ooligan Press.

Hou, Jeffrey. 2010. *Insurgent public space: guerrilla urbanism and the remaking of contemporary cities*. New York: Routledge.

Iveson, Kurt. 2010. "Introduction: Graffiti, street art and the city." *City* 14 (1/2): 25–32.

Jacobs, Jane. 1961. *The death and life of great American cities*. New York: Random House.

Kelling, George L and Catherine M Coles. 1996. *Fixing broken windows: restoring order and reducing crime in our communities*. New York: Martin Kessler Books.

Lofland, John. 1998. *The public realm: exploring the city's quintessential social territory*. Hawthorne, NY: Aldine de Gruyter.

Lopez, Adriana. 2013. "New Orleans Jazz Fest comes full circle with its mission." www.forbes.com/sites/adrianalopez/2013/05/06/new-orleans-jazz-fest-comes-full-circle-with-its-mission

Loukaitou-Sideris, Anastasia and Renia Ehrenfeucht. 2009. *Sidewalks: conflict and negotiation over public space*. Cambridge, MA: MIT Press.

Lovett, Ian. 2013. Food carts in Los Angeles come out of the shadows. *New York Times* December 4.

Low, Setha, Dana Taplin, and Suzanne Scheld. 2005. *Rethinking urban parks: public space and cultural diversity*. Austin, TX: University of Texas Press.

Markusen, Anne and Anne Gadwa. 2010. *Creative placemaking*. Washington, DC: National Endowment for the Arts.

Markusen, Ann and Greg Schrock. 2009. "Consumption-driven urban development." *Urban Geography* 30(4): 344–367.

Merker, Blaine. 2010. "Taking place: Rebar's absurd tactics in generous urbanism." In *Insurgent public space: guerrilla urbanism and the remaking of contemporary cities*, edited by Jeffrey Hou. New York: Routledge.

Morales, Alfonso. 2000. "Peddling policy: street vending in historical and contemporary contest." *International Journal of Sociology and Social Policy* 20(3/4): 76–98.

Morales, Alfonso. 2010. "Planning and the self-organization of marketplaces." *Journal of Planning Education and Research* 30(2): 182–197.

Morales, Alfonso and Gregg Kettles. 2009. "Healthy food outside: farmers' markets, taco trucks, and sidewalk fruit vendors." *Journal of Contemporary Health Law & Policy* 26(1): 20–48.

Mukhija, Vinit and Anastasia Loukaitou-Sideris. 2014. *The informal American city: from taco trucks to day labor*. Cambridge: MIT Press.

Mural Arts Program. n.d. "City of Philadelphia Mural Arts Program." www.muralarts.org

Murdoch, James, Carl Grodach, and Nicole Foster 2015. "The importance of neighborhood context in arts-led development: community anchor or creative class magnet?" *Journal of Planning Education and Research* online August 18. doi: 10.1177/0739456X15599040

NASAA. 2013. *State Policy Briefs: Percent for Art*. Washington, DC: National Assembly of State Arts Agencies. www.nasaa-arts.org/Research/Key-Topics/Public-Art/NASAAPercentforArtPolicyBrief.pdf

Newman, Lenore Lauri and Katherine Burnett. 2012. "Street food and vibrant urban spaces: lessons from Portland, Oregon." *Local Environment* 18(2): 233–248. doi: 10.1080/13549839.2012.729572

Nicodemus, Anne Gadwa. 2013. "Fuzzy vibrancy: creative placemaking as ascendant US cultural policy." *Cultural Trends* 22(3/4): 213–222.

Olsen, Donald J. 1986. *The city as a work of art: London, Paris, Vienna*. New Haven, CT: Yale University Press.

Prou, Sybille and King Adz. 2008. *Blek le Rat: getting through the walls*. New York: Thames & Hudson.

Reiss, J. 2007. *Bomb it: Street art is revolution* (documentary film). Flying Cow Production.

Sharp, Joanne, Venda Pollock, and Ronan Paddison. 2005. "Just art for a just city: public art and social inclusion in urban regeneration." *Urban Studies* 42(5/6): 1001–1023.

Shouse, Heather. 2011. *Food trucks: dispatches and recipes from the best kitchens on wheels*. New York: Ten Speed Press.

Stern, Mark J and Susan C. Seifert. 2010. "Cultural clusters: the implications of cultural assets agglomeration for neighborhood revitalization." *Journal of Planning Education and Research* 29(3): 262–279.

Stevens, Quentin. 2007. *The ludic city: exploring the potential of public spaces*. London: Routledge.

Urban Conservancy and Civic Economics. 2009. *Thinking outside the box*. New Orleans, LA: Urban Conservancy and Austin, TX: Civic Economics. www.staylocal.org/sites/default/files/reports/ThinkingOutsidetheBox_1.pdf

USDA. 2013. "USDA celebrates National Farmers Market Week, August 4–10." US Department of Agriculture. www.usda.gov/wps/portal/usda/usdamediafb?contentid=2013/08/0155. xml&printable=true&contentidonly=true

USDA. 2014. "New data reflects the continued demand for farmers markets." US Department of Agriculture. www.usda.gov/wps/portal/usda/usdahome?contentid=2014/08/0167.xml

Vitale, Alex S. 2008. *City of disorder: how the quality of life campaign transformed New York politics.* New York: NYU Press.

Wessel, Ginette. 2012. "From place to NonPlace: a case study of social media and contemporary food trucks." *Journal of Urban Design* 17(4): 511–531.

Whybrow, Nicholas. 2011. *Art and the city.* London: I.B. Tauris.

Whyte, William Hollingsworth. 1980. *The social life of small urban spaces.* Washington, DC: Conservation Foundation.

Whyte, William Hollingsworth. 1988. *City: rediscovering the center.* New York: Doubleday.

Witz, Dan. 2010. *Dan Witz: in plain view: 30 years of artworks illegal and otherwise.* Berkeley, CA: Gingko Press.

Zukin, Sharon. 2010. *Naked city: the death and life of authentic urban places.* Oxford: Oxford University Press.

SECTION IV

Urban Revitalization Methods

Data, Techniques, and Relationships

The previous chapters have focused on developing a critical understanding of urban revitalization theory, history, and policy-making in different contexts. Understanding the successes and challenges of urban revitalization programs in different times and places is important to inform future decision-making. However, it is not enough. We also benefit from studying the unique conditions and specific context where work will take place, and from engaging community members in the planning process. In the next three chapters, we focus on how to build different forms of knowledge and skills necessary to negotiate revitalization planning processes. In chapter 12 we introduce readers to the important data sources and analytical techniques that are useful to understanding various facets of community life and how neighborhoods change over time. This serves as a basis for developing strategic urban revitalization policy goals. Next, in chapter 13, we focus on field methods such as site observation techniques, surveys, and ethnographic methods used to understand neighborhood conditions as well as residents' perspectives and experiential information about a community. Finally, chapter 14 delves into community participation and engagement techniques for revitalization planning alongside a brief discussion of intergovernmental relations and the institutional context of planning and development.

12

Data sources and community assessment tools

A first step in any revitalization planning process is to establish a baseline understanding of the economic, physical, and social conditions that characterize a community. Population demographics, resident occupations, homeownership rates, types of neighborhood businesses and services all matter in shaping an urban revitalization program. Is the area facing a shortage of affordable housing? Is crime a problem in specific areas? Is there a shortage of access to healthy foods? We can gather public data to answer many of these questions and go into the urban revitalization planning process with informed knowledge of the community. Whether we are working in a shrinking city, an inner suburban neighborhood, or a gentrifying central city area, understanding a community's characteristics is crucial. This knowledge allows us to identify and better understand opportunity areas and important challenges and, ultimately, set an agenda for urban revitalization.

In this chapter, we introduce key public data sources and analytical techniques for urban revitalization planning in these different contexts. We present and discuss relevant demographic, economic, and consumer data that both professionals and community groups can use to better understand the unique conditions in a neighborhood and its surrounding region. We also explain how to use these data sources to create a community profile and market analysis. By the end of the chapter, you will know how to gather data to build a profile of the local population, workforce, and local businesses; define a community's trade area; develop indicators to monitor and measure community change; and conduct a basic retail gap analysis. This information is essential to establish an understanding of a community and to identify patterns of change in relation to the larger city and region. When used in conjunction with the field methods and community participation techniques discussed in the preceding chapters, we can develop deep knowledge of local conditions to inform an urban revitalization plan.

GATHERING AND UNDERSTANDING DATA FOR URBAN REVITALIZATION

The aim of this section is to help readers learn to navigate and analyze the array of public data available for urban revitalization planning. The US government provides a wealth of free, publicly accessible data that we can use to help understand urban areas at a variety of scales, from a city block to the entire metropolitan area. The data help develop better knowledge about

TABLE 12.1 KEY PUBLIC DATA SOURCES

Source	Type of data
American Community Survey (ACS), US Census	Residential demographics and housing
Consumer Expenditure Survey (CES), Bureau of Labor Statistics	Consumer spending
County Business Patterns (CBP), US Census	Business and industry
Occupational Employment Statistics (OES), Bureau of Labor Statistics	Workforce and occupations
Department of Housing and Urban Development (HUD)	HUD subsidies, subsidized households, foreclosure and vacancy rates
Home Mortgage Disclosure Act (HMDA)	Mortgage lending statistics

the residential population including demographic characteristics, housing and neighborhood conditions, employment, purchasing power, and preferences. Demographics refer to people's characteristics such as age, income, and household structure. Purchasing power is the amount of money that a given household or community has to spend on particular goods or services. Preferences describe the different spending and life patterns among people. We can acquire data to analyze these community features and assess conditions at a point in time or analyze change over a period of time. This knowledge helps to identify potential revitalization strategies appropriate to unique local conditions.

Table 12.1 presents the essential sources of public data for urban revitalization planning that we discuss below.

AMERICAN COMMUNITY SURVEY

The primary source of residential data nationwide is the American Community Survey (ACS). This is a good starting point for understanding the social conditions and economic opportunities in a community because it provides valuable information on residential demographics and housing characteristics. Looking at demographic data over time helps explain how a neighborhood or city is changing based on key features including level of education, racial composition, the occupation of residents, or how far they travel to work.

Until 2010, the decennial Census consisted of both a "short form," which provided basic demographic information, and a "long form," used to gather more detailed information on residential and housing data. Beginning in 2010, the ACS replaced the Census long form as the source of detailed and timely information on community characteristics. While both provide population information, they are not necessarily comparable. The Census provides a series of useful handbooks available online that provide further details on how to use ACS data (US Census Bureau n.d.-a).

We can use the ACS to collect information on household characteristics (one or more persons living in the same housing unit) in a community as well as the absolute number of residents there. Households may be categorized by size or composition (family or nonfamily members living together). Because households often make similar decisions, we collect most of the important data at this level. Table 12.2 shows the residential data available from the ACS that are important to collect for a community analysis.

Using the ACS can be tricky, in part because data are reported based on three different estimates (one, three, and five years). Each has different strengths and weaknesses and is used in different

TABLE 12.2 KEY ACS RESIDENTIAL DATA CATEGORIES

Population characteristics	Economic characteristics	Social characteristics	Housing characteristics
Absolute population Number of households Household size and composition Gender Age	Average or median household income Income per capita Percentage of population living in poverty Employment status (employment and unemployment rates) Occupation and industry participation Journey or travel time to work	Race and ethnicity Place of birth Language spoken at home Education level	Housing tenure (owner- or renter-occupied units) Year structure built Units in structure

contexts depending on what you need to know and the geographic scale that you are studying. The one-year ACS is based on 12 months of data collection. This is the most current source, but also is the least reliable because it has a comparatively low sample rate. The three-year ACS and five-year ACS are based on 36 and 60 months of data collection, respectively. They are more reliable due to their larger sample sizes, but do not have the currency of the one-year estimates because the data are collected over multiple years. Another challenge comes when comparing ACS data across multiple years. Because the multi-year estimates are collected over time rather than in one year, we cannot compare easily from one period to another. While the data gathered from the multi-year estimates are technically not comparable, some researchers use the midpoint in the data as an estimate for time-series comparison. For example, 2009–13 ACS data may be considered to represent 2011. In sum, for the most accurate and reliable results we must uniformly collect data from one of the three estimates and, if using the multi-year estimates, take care in comparing data over time.

Another caution concerns the geographies at which the Census reports ACS data. All ACS data are not uniformly available at every geographic level. The ACS provides thorough coverage for virtually all metropolitan statistical areas (MSAs) across each estimate, but this is not the case for smaller geographies. In fact, studies at the neighborhood level must rely on five-year estimates because this is the only source for zip code, census tract, and block group information. When comparing metro- and neighborhood-level data, be sure to use the five-year estimate so that you know your results will be comparable.

CONSUMER EXPENDITURE SURVEY

The Consumer Expenditure Survey (CES) from the Bureau of Labor Statistics is an underutilized but extremely useful source of data in urban planning and revitalization. The CES provides annual data on household spending across a wide variety of categories from basic daily needs such as groceries and transportation to items including clothing, entertainment, and health care. The CES is a crucial tool to estimate how different households spend their income based on different demographic characteristics and geographies. Using the CES in conjunction with ACS data, we can estimate the total and annual spending in a community or its purchasing power in a variety of categories. We can also compare this to the rest of the US and determine if an area is under- or overrepresented in a particular retail niche or everyday service. Community purchasing power is particularly important for urban revitalization initiatives because high-density poor communities often compensate for their lower average income. Demonstrating an area's purchasing power can make a compelling case to attract retailers that may improve the quality of life in lower-income communities.

The Bureau of Labor Statistics provides data for various demographic characteristics that allows the analyst to produce a more nuanced view based on the demographics of their neighborhood or region. This allows you to fine-tune the data further by creating categories based on the specific socio-economic characteristics of your study area. This is what private data companies do to produce their tailored market segmentation reports such as ESRI's Tapestry, or Claritas Prizm. The Bureau of Labor Statistics also releases estimates by US regions and major metropolitan areas (e.g. Los Angeles, New York, Dallas–Fort Worth, or Baltimore).

BUSINESS PATTERNS

Business Patterns data from the US Census are an essential resource to help understand the characteristics of area businesses. The data are based on surveys of businesses and administrative records from the Internal Revenue Service among other sources. Unlike the ACS, Business Patterns data are based on firm location and include data on employees and their workplace. Whereas the ACS tells you about residents in your trade area, and CES about their spending habits, Business Patterns data tell you about the industries and businesses.

Business Patterns data are organized under the North American Industrial Classification System (NAICS). The NAICS is a hierarchical system that codes business activity based on industry classification. Data are organized under a set of 20 two-digit industry codes that increase specificity up to six-digit codes. Two-digit industries capture basic economic activity such as Wholesale trade (42) or Accommodation and food service (code 72). The six-digit codes within each of these offer considerably more specificity, such as Household appliances, electric housewares and consumer electronics merchant wholesalers (423620), or Caterers (722320). Based on each code, NAICS provides data on total employment, payroll, and the total number and size of establishments. The data are available at five scales: US, state, metropolitan area, county, and zip code. In most cases, Zip Code Business Patterns is the most appropriate dataset to understand neighborhood economic activity, while county and metropolitan area data are useful to understand regional economic activity and contextualize a local revitalization study.

There are some limitations and challenges to using these data. First, NAICS does not capture all economic activity. The NAICS does not include agricultural production, most government employment, or self-employment. The Census does report self-employment separately in Nonemployer Statistics. This comprises about 11% of all US employment, but it accounts for very little economic activity. As a result, while the omitted industries will likely have little effect on understanding the regional economy, those places with robust informal economies will be undercounted. We can get a sense of this activity, however, through other field methods (chapter 13). Second, another issue is that some data are suppressed or not directly reported because it may compromise confidentiality requirements. In this case, the Census reports a range of employment (e.g. 0–19 employees, 20–99 employees, etc.). The most basic technique to estimate suppressed employment is simply to use the mean in the range provided (e.g. 59.5 or 60 for 20–99 employees). Finally, it is very difficult to examine industry change over an extended period of time. NAICS codes have only been in use since 1997. Prior to this, the Census classified industry activity under the Standard Industrial Classification (SIC) system. The NAICS is a significant improvement over the SIC because it captures the shift in our economy towards services and better reflects high-technology industries, but the two coding systems are not directly comparable. While the Census provides a crosswalk to compare the different codes, many important industries do not have an exact match, which makes comparison prior to 1998 next to impossible for these industries.

In addition to Business Patterns data, other government business data sources include the Economic Census, Quarterly Census of Employment and Wages (QCEW), Survey of Business Owners, Quarterly Workforce Indicators (QWI), and the abovementioned Nonemployer Statistics. However, most of these sources do not contain data below the county or regional level, and some provide data only twice during a decade (in years ending in "2" and "7"). They can, nonetheless, provide important contextual information for many analyses.

OCCUPATIONAL EMPLOYMENT STATISTICS

Occupational Employment Statistics (OES), produced by the Bureau of Labor Statistics, provide employment and wage data for occupations within and across industries. The data are organized under the Standard Occupational Classification (SOC) system, which organizes occupations into 23 major groups and 840 specific occupations. OES data enable us to create occupational profiles based on both industries and places, and therefore they complement the NAICS data because they provide a picture of the workforce skill base. While the NAICS provides total employment counts irrespective of job title or responsibilities, using SOC data we can determine what people actually do in an industry, and the percentage of occupations that comprise an industry's employment base. For example, using SOC data we can determine that in Book, Periodical and Music Stores (NAICS 451200), 81% of employees are involved in sales and related activities, 2.27% in management, and fewer than 1% are employed in arts and design occupations (US Bureau of Labor Statistics 2011). This information is useful when considering the types of jobs you want to bring to a community, but keep in mind that these industry-specific breakdowns are based on national-level data and not individual communities. The OES also report employment and wage data for occupations at the state and metro area levels and provide details on the places with the highest employment level, concentration, and pay for each occupation. The OES program does not provide easily accessible data below the metro level and, like Business Patterns data, does not provide self-employment data. When it becomes necessary, more advanced researchers can access this information using the Public Use Microdata Sample (PUMS).

Beyond these sources, there are a variety of data sources available to help understand neighborhood housing conditions and local and regional housing markets.

DEPARTMENT OF HOUSING AND URBAN DEVELOPMENT

The Department of Housing and Urban Development (HUD) provides access to data on foreclosures and vacancy rates (from the US Postal Service) and on housing choice voucher, public housing funding, and Low Income Housing Tax Credit (LIHTC) activity. These data can answer many questions about neighborhood conditions. For example, how many properties have been foreclosed in a neighborhood over a given time period and how does this rate compare to the regional average? How many properties are vacant for 12 months or more? Where is LIHTC-funded housing located in a region? In addition to these data, HUD's Picture of Subsidized Households (PSH) provides data on characteristics of households that receive assistance through HUD programs. This can be an important source of data to learn specifically about the demographic composition and potential employment and education gaps in these settings, and they can be compared to the neighborhood as a whole using ACS data.

HOME MORTGAGE DISCLOSURE ACT

Home Mortgage Disclosure Act (HMDA) data provide information on residential lending markets. This source gathers data on the volume, number, and type of mortgages made (new home purchase, home improvement, or refinance) down to the level of the census tract, as well as the number of mortgage applications that were denied in a given year and place. The County Assessors' Office can provide data on home sales and property values, and local sources are available for building permit data housing and housing code violations.

OTHER SOURCES

Other potential sources of community data include education data from local school records and the National Center for Education Statistics on attendance, enrollment, test performance grades, and student-to-staff ratios. Local police records can provide public safety information on violent and property crimes. Finally, some private companies provide free community data, such as Walkscore, which rates neighborhood walkability.

CHALLENGES

It is important to note that public data present some challenges. First, we must decide what data are most relevant to collect for a given community. The US Census alone reports on hundreds of possible topics to explore in nearly a dozen different public data sources. Second, these data sources do not always provide comparable data across different years and geographies. Third, once we do decide the appropriate data sources, years, and areas of analysis, we must figure out how to organize the data in a meaningful way for analysis and presentation. Finally, virtually all sources report data based on samples and, as a result, there is a degree of uncertainty about whether the sample estimates reflect the population. Estimates are used because counting every single individual or business and collecting detailed information from all of them on an annual basis would be a monumental task. Even the US Census, the ten-year count of every person living in the United States, has limits. The limits are particularly problematic for communities in need of revitalization because the Census may undercount some low-income and minority populations (El Nasser and Overberg 2012). Because the Census underreports their numbers—and in some cases income—it may inaccurately deflate an area's potential purchasing power since most analyses rely on Census data. This in turn can influence retailer interest in an area, as we discussed in chapter 7. Despite these shortcomings and challenges, public data sources provide very useful and accessible sources of comparative data for urban revitalization planning.

There are also a number of valuable commercial sources available. Private companies such as ESRI and Claritas allow users to produce reports based on a combination of data sources. This takes out much of the legwork required to access and compile public data, but users must pay an annual fee for access to the data, and generate reports. Small cities and community organizations often do not possess a budget to maintain subscriptions to these data. For this reason, many revitalization professionals use publicly available data.

THE COMMUNITY TRADE AREA

Before gathering data, we must first define an area of analysis. One way to do this is by defining a community trade area. The trade area is essentially the area from which businesses in a community draw their customers. This primarily encompasses the targeted community and the surrounding area where residents spend most of their time and money, but also we consider that local businesses may draw customers from beyond the immediate community. The trade area is based foremost on geography and travel distance. We also take into consideration the backgrounds, preferences, demand, and purchasing power of residents and customers in the surrounding community and competing areas. We use these data to analyze the area's strengths and weaknesses, identify retail opportunities, and build strategies toward enhancing economic opportunity in a given area.

To define the trade area, researchers must make assumptions about its size, location, and characteristics. While we could use the boundaries of a neighborhood to define the trade area, this approach risks overlooking important information about the specific conditions there. One important consideration is that different types of business have different trade areas. For example, people generally will travel a relatively long distance to attend arts and sporting events, or to shop for specialty items like furniture or automobiles. Such activities tend to cluster together, creating destination points. This enables consumers to compare the price and quality of similar items. Conversely, people need immediate access to groceries or restaurants and everyday retail such as a small hardware store or drycleaner. These different types of goods and services can be grouped into different types of trade areas comprised of "comparison" and "convenience" goods. Whereas comparison goods will draw customers from throughout the region and sometimes beyond, convenience goods tend to depend on their surrounding neighborhoods and serve the immediate population. As a result, they will have very different trade areas. In general terms, neighborhood retail trade areas will encompass an area between 0.5 and 3 miles and serve 3,000–30,000 people (Jacobus and Hickey 2006). Comparison goods in regional shopping centers will have a larger trade area that encompasses 10–12 miles (Gibbs 2007).

The reality, however, is that every place is comprised of a myriad of overlapping trade areas that vary with the size, variety, and density of the business mix, characteristics of the built and natural environments, and local demographics. Natural barriers such as lakes or rivers, road capacity, and population density all affect the size and shape of a trade area, as do competing businesses in nearby commercial districts. Trade areas can vary by time of day too. Business districts, for instance, will have an active daytime trade area that usually disappears at night and on weekends. The opposite may be said for entertainment areas that attract most people in the evenings or for special events. In both instances, paying attention to time can help to identify business opportunities during nonpeak hours or address issues such as perceptions of safety at night.

The presence of an anchor that draws a large number of people is also an important factor. Anchors can be destinations like department stores. Libraries, universities, hospitals, museums, and even parks also serve as anchors that attract people to an area. Spectacular places discussed in chapter 6, like the High Line in New York or Millennium Park in Chicago, are good examples, but so too are smaller, more modest public places that anchor neighborhoods, such as Rittenhouse Square in Philadelphia or Washington Square in San Francisco. Alternatively, dense retail areas comprised of small businesses can be destinations in their own right. This is not only the case in older downtowns, but also occurs in suburban areas such as Little Saigon (see Box 8.2). These areas can have larger trade areas because the density of activity allows customers to combine activities and because customers are drawn to the uniqueness of the area.

Finally, demographic features can affect trade area characteristics. For example, a study of retail activity in New York found that low-income neighborhoods have a smaller, less diverse retail mix and fewer chain stores and restaurants than wealthier areas. Latino neighborhoods have better access to retail and food services than predominantly white neighborhoods (Meltzer and Schuetz 2011). In short, when defining the trade area it is important to consider not only physical and economic features, but also the characteristics of customers and residents.

With all these different variables, how do you go about defining a trade area? There are a variety of different methods that make it relatively easy to gather data and prepare a community profile. Professionals often conduct focus groups or surveys with customers and residents in the area targeted for revitalization to document where they live and shop (see chapter 13). Businesses may also collect customer records allowing one to map where customers live (based on street address or zip code) and using the information from multiple representative businesses to produce a trade area based on the area where the majority of customers live. The challenge here is obtaining a representative sample of customers across different business types.

Other approaches rely on GIS or similar software to map data onto an area. For example, the simplest approach is to draw concentric rings (e.g. one mile, three miles, five miles) from the midpoint of the area and collect data based on these different-sized trade areas. This method provides a basic understanding of the trade area, but does not consider barriers to travel or competition from adjacent trade areas.

A similar but more sophisticated approach is to produce a drive-time analysis. This approach uses shapes or polygons of travel time and distance (e.g. ten-minute drive, 20-minute walk) to calculate the trade area. You can conduct this analysis using GIS or with free software available online such as freemaptools.com or Google Maps. A ten-minute drive time is a common basis to define the general trade area, and estimates usually range from about five to ten minutes for neighborhood retail to 15 minutes for a larger shopping center, and up to 30 minutes for a mall. The limitation of this approach is that it does not take traffic time into account. You also must manually account for neighboring competition (Figure 12.1). With both the drive-time and concentric ring approaches, you can use ESRI Business Analysis or GIS to produce reports that include estimates based on the rings or polygons, or you can approximate the area yourself based on zip codes or census tracts.

While other more sophisticated models exist that attempt to predict customer behavior based on competition and shopping patterns, the models are only as good as the data going into them. The reality is that defining the trade area is not an exact science, but more of an art based on a combination of local knowledge and available data. Given the variability and difficulty of identifying the trade area, planners will often identify and compare data from two or even three trade areas and adjust their survey or drive-time data based on known competition.

Additionally, it is typical to compare your trade area to two or more areas with similar characteristics based on criteria such as population size, demographics, and regional location. These can be areas you view as being in direct competition with your area, or similar places that have undergone revitalization that the community may aspire to emulate.

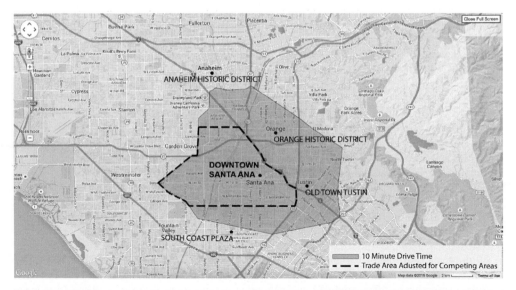

Figure 12.1   Defining a community trade area for Downtown Santa Ana (based on a 10 minute drive time and adjusted for local competition)

Source: www.freemaptools.com and Google Maps

COMMUNITY ASSESSMENT INDICATORS AND TOOLS

Once we have gathered descriptive data on the community, we can employ this information to monitor and measure community change. Indicators clue us in to community strengths and weaknesses and enable us to determine if we are meeting revitalization goals. For example, some studies look to housing foreclosures as a warning sign of neighborhood decline, while others have attempted to develop a wider set of "neighborhood distress" indicators based on social and economic variables including poverty, unemployment, income, education levels, and foreclosure rates (Jennings 2012; Williams, Galster, and Verma 2013). Similarly, the Urban Institute's National Neighborhood Indicators Partnership gathers a wide range of data to "build the capacities of institutions and residents in distressed urban neighborhoods" (NNIP n.d.). Collecting community indicators is one route toward helping us determine if we need to focus efforts, for example, on housing affordability, increasing employment training opportunities, or addressing a dearth of neighborhood retail. They are also being used to compare and monitor outcomes for complex programs such as Choice Neighborhoods and Promise Neighborhoods (Smith 2011).

In terms of economic revitalization, we can apply basic analytical techniques to determine the relative concentration of different types of business and employment in the community. We can identify strong sectors and consumer market segments as well as weaker areas. Additionally, we can determine the surplus or leakage of resident consumer spending in the area. These indicators, particularly when used in comparison with other areas, will help to answer questions important in understanding neighborhood opportunities and challenges, such as: What are the strongest types and levels of employment in the trade area? What types of business are most concentrated in the community? Are these strengths unique to the community when compared to other, similar places? In what types of retail activity is the community underserved? How much money do residents spend outside the community? Below, we discuss techniques that will help provide insight into area strengths, weaknesses, and opportunities, and point toward strategies for future development. Box 12.1 lists some of the key terms that will be explained in the following sections.

BOX 12.1 COMMUNITY ASSESSMENT KEY TERMS

- Business inventory
- Business mix
- Capture rate
- Community trade area
- Drive-time analysis
- Gap analysis
- Location quotient
- Purchasing power
- Sales leakage
- Sales surplus

LOCATION QUOTIENT

The location quotient (LQ) is a common and useful technique in economic development, used to gauge how specialized or concentrated a particular type of industry is in the local economy. The LQ is formulated based on the ratio between the percentage of employment in a local industry to that of the same industry in a larger reference economy such as the state or nation. The LQ formula is:

LQ = (local employment in industry / total local employment) / (industry employment in reference economy / total employment in reference economy)

We interpret an LQ of 1 to mean that the local economy has the same level of concentration as the larger reference economy. An LQ less than 1 indicates a weakness in the local industry or that the local economy is underrepresented compared to the reference economy. An LQ greater than 1 indicates a specialization in that industry. An LQ of 1.25 or greater represents a strong local presence. Although economic developers most commonly use LQs to measure industry employment concentrations, we can also use the formula to determine the concentrations based on occupations or business establishments.

Because labor markets tend to be regional, we typically use the LQ to measure industry concentration at the regional level, and use the nation as the reference economy. For urban revitalization analyses, this is helpful to gain insight into the regional economic context. Although some cities will use municipal boundaries to define the local economy, this approach misses industry concentrations important to the city in the surrounding jurisdictions.

At a smaller scale, we can use the LQ to get a general sense of employment specialization in a neighborhood using zip codes as a proxy. In this case, it is best to use a smaller reference economy such as the state or metropolitan area since you want to determine how your trade area compares to the regional economic mix. However, keep in mind that the LQ is not an ideal measure for smaller geographies because they typically contain a less diverse industry mix than the regional economy, and you might miss activity in immediately adjacent areas. You can partially address this by comparing different-sized trade areas. Keep in mind that areas with very low employment, but with a few highly specialized employers, will produce very high LQs, so it is always useful to compare the LQ to the absolute level of employment. An alternative measure is per capita employment (the ratio of employment in area businesses to the local population). This measure will inflate places with high employment and low population.

THE BUSINESS MIX

Another important technique is to document the business mix in a community. To analyze the business mix, we need to produce an inventory of the existing businesses within the trade area, organize them by NAICS code, and evaluate the trade area against comparison areas (Table 12.3). You can gather data using Zip Code Business Patterns or the field methods discussed in chapter 13. The business inventory allows the analyst to understand the local business landscape and identify any anchor businesses such as a department store or a major supermarket. Additionally, by classifying businesses in the inventory by percentage and evaluating them against comparison areas, you can identify potential retail clusters or market niches (e.g. concentrations of specialized or complementary services such as auto dealerships, restaurants, antique stores, or ethnic products) as well as retail services that are potentially undersupplied. The LQ is helpful to get a sense of the community market supply as well, because it allows you to measure the concentration of retail establishments compared to a larger reference economy although, again, you should keep in mind that the LQ will not account for activity in adjacent areas.

TABLE 12.3 BUSINESS INVENTORY (RETAIL TRADE)

NAICS	Business description	Trade area		Comparison area 1		Comparison area 2	
		Count	Percentage	Count	Percentage	Count	Percentage
441110	New car dealers	2	1.1	5	2.2	0	0
441120	Used car dealers	0	C.0	2	0.9	0	0
441310	Automotive parts and accessories stores	1	0.6	6	2.7	0	0
441320	Tire dealers	0	0	2	0.9	0	0
442110	Furniture stores	0	0	3	1.3	0	0
442210	Floor covering stores	0	0	2	0.9	0	0
442299	All other home furnishings stores	3	1.7	4	1.8	4	3.1
443111	Household appliance stores	0	0	3	1.3	0	0
443112	Radio, television, and other electronics stores	2	1.1	17	7.6	12	9.4
443120	Computer and software stores	0	0	1	0.4	1	0.8
444120	Paint and wallpaper stores	0	0	2	0.9	0	0
444130	Hardware stores	0	0	3	1.3	2	1.6
444190	Other building material dealers	1	0.6	5	2.2	2	1.6
445110	Supermarkets and other grocery (except convenience) stores	4	2.3	23	10.3	7	5.5
445120	Convenience stores	16	9.0	3	1.3	3	2.4
445210	Meat markets	3	1.7	2	0.9	1	0.8
445220	Fish and seafood markets	2	1.1	0	0	0	0
445230	Fruit and vegetable markets	0	0	2	0.9	3	2.4
445291	Baked goods stores	2	1.1	8	3.6	4	3.1
445299	All other specialty food stores	2	1.1	4	1.8	2	1.6
445310	Beer, wine, and liquor stores	1	0.6	6	2.7	2	1.6
446110	Pharmacies and drug stores	12	6.8	17	7.6	4	3.1
446120	Cosmetics, beauty supplies, and perfume stores	1	0.6	12	5.4	4	3.1
446130	Optical goods stores	0	0	3	1.3	1	0.8
446191	Food (health) supplement stores	5	2.8	7	3.1	0	0
446199	All other health and personal care stores	0	0	5	2.2	0	0
447110	Gasoline stations with convenience stores	2	1.1	6	2.7	0	0

TABLE 12.3 continued

NAICS	Business description	Trade area		Comparison area 1		Comparison area 2	
		Count	Percentage	Count	Percentage	Count	Percentage
448110	Men's clothing stores	1	0.6	2	0.9	3	2.4
448120	Women's clothing stores	11	6.2	9	4.0	7	5.5
448130	Children's and infants' clothing stores	1	0.6	1	0.4	1	0.8
448140	Family clothing stores	2	1.1	4	1.8	5	3.9
448150	Clothing accessories stores	3	1.7	2	0.9	3	2.4
448190	Other clothing stores	1	0.6	2	0.9	4	3.1
448210	Shoe stores	2	1.1	5	2.2	0	0
448310	Jewelry stores	26	14.7	6	2.7	8	6.3
448320	Luggage and leather goods stores	1	0.6	1	0.4	0	0.0
451110	Sporting goods stores	1	0.6	3	1.3	4	3.1
451120	Hobby, toy, and game stores	2	1.1	1	0.4	2	1.6
451130	Sewing, needlework, and piece goods stores	0	0	1	0.4	0	0
451140	Musical instrument and supplies stores	1	0.6	4	1.8	0	0
451211	Book stores	6	3.4	6	2.7	0	0
451212	News dealers and newsstands	2	1.1	0	0	0	0
451220	Prerecorded tape, compact disc, and record stores	3	1.7	0	0	1	0.8
452990	All other general merchandise stores	7	4.0	5	2.2	5	3.9
453110	Florists	1	0.6	3	1.3	1	0.8
453210	Office supplies and stationery stores	2	1.1	1	0.4	2	1.6
453220	Gift, novelty, and souvenir stores	29	16.4	2	0.9	11	8.7
453310	Used merchandise stores	3	1.7	3	1.3	1	0.8
453910	Pet and pet supplies stores	1	0.6	1	0.4	1	0.8
453998	All other miscellaneous store retailers (except tobacco stores)	4	2.3	3	1.3	7	5.5
454111	Electronic shopping	4	2.3	4	1.8	9	7.1
454113	Mail-order houses	2	1.1	1	0.4	0	0
	Total	177		223		127	

It is also useful to obtain more specific data on businesses in the trade area. This includes estimates of the gross leasable area (GLA) or total square footage of building space dedicated to different retail activities, total sales, parking availability, and traffic counts for retail cluster or corridor locations. These data may come from a business inventory, survey, or national sales estimates published by the Urban Land Institute's *Dollars & cents of shopping centers* (2008) or the *Retail tenant directory* (Trade Dimensions n.d.). You can also speak with commercial real estate agents in the area. This can be used later to compare with the location selection criteria of major retailers or to estimate retail gaps and community purchasing power, which we discuss below.

GAP ANALYSIS

Gap analysis is a technique that goes beyond a simple comparison of businesses in the trade and comparison areas to calculate the extent to which the current business mix meets residential consumer demand. This technique compares the community's purchasing power or overall spending (demand) to the sales by businesses in the trade area in order to estimate a local sales surplus or leakage. This allows us to estimate how much residents spend outside the community (sales leakage) or determine when sales exceed local spending (sales surplus). This information helps to identify more precisely retail clusters, segments of underserved retail, and the capacity for additional retail activity in the area.

To conduct a gap analysis, you need three sources of data: ACS population and income data, CES data, and sales data (from Urban Land Institute 2008 or your business inventory). The next steps involve estimating potential spending and sales. The basic gap formula to figure the sales surplus/leakage is:

surplus/leakage = estimate of total potential spending − estimate of total sales by retail category

To estimate resident spending in the trade area requires data on trade area demographics and spending patterns. At a minimum you will want to know the total household population in the trade area (based on the ACS) and how much households spend on the retail sector of interest (based on the CES). However, you can produce a more accurate estimate by calculating a spending estimate based on population characteristics such as median household income. In this case, the formula to estimate area spending on groceries (labeled food at home in the CES) would be:

trade area grocery spending = household population × median household income × percentage of total household spending on food at home

This gives you an estimate of potential spending because it is based on the assumption that residents only shop in the trade area for goods and services. In addition to household income, you can further refine your estimate by consulting various permutations of spending in the CES (based on combinations of income and age, race or homeownership status, for instance) that most closely match the demographics of the trade area. Nuanced analyses will take into account that while higher-income households spend more on food, lower-income households spend a higher percentage of their income on food, or homeowners spend more on household furnishings.

The next step is to estimate total sales in the trade area. The most accessible source for these data is Urban Land Institute (2008), which provides national estimates of total sales per square foot for hundreds of different retail activities. You can acquire more specific data or augment the national estimates by conducting a business inventory. With these data, one can estimate the total square footage of trade area stores for the category of interest by the estimate of sales per square foot.

Comparing total spending (or potential resident spending) to total sales in each retail category gives you the sales gap or surplus. If total spending is less than total sales, the area contains a sales surplus. This implies that people are traveling from outside the area to consume trade area goods and services. Conversely, if total spending exceeds total sales, the area contains a sales gap or sales leakage. This implies that residents are spending their income on businesses outside the trade area. Once you have the data on sales (supply) and spending (demand) you can simply determine the sales leakage (gap) or surplus using the formula above.

From there, the gap analysis also allows you to estimate the existing capture rate or percentage of resident spending in the community. Calculate the capture rate by dividing trade area spending by area sales:

capture rate = trade area sales / total spending

Based on the existing capture rate and consideration of location characteristics such as competition from adjacent areas, available land, and available building space, we can make an assumption on the potential capture rate or additional market share that the community can support.

Box 12.2 provides a guide through the process of figuring the grocery store sales gap and capture rate for a hypothetical trade area. In sum, the gap analysis allows us to determine if a sales gap exists in a local retail sector and provides a basis to determine the market potential for filling that gap based on the community's purchasing power.

BOX 12.2 GROCERY STORE SALES GAP ANALYSIS

Zip code ———
Households: **14,256**
Median household income: **$38,463**
Retail category: **Supermarkets and other grocery (except convenience) stores (NAICS 445110)**
Total stores in trade area: **two establishments, 55,000 total square feet (20,000 and 35,000 square feet each)**
Grocery sales per square foot: **$390.25**
Expenditure share on food at home ($30,000–39,999 median household income): **8.9%**
Trade area spending on groceries (demand potential):
14,256 × $38,463 × 8.9% = $48,801,239
Trade area grocery sales (supply): **55,000 square feet × $390.25 sales per square foot = $21,463,750**
Results:
Spending estimate (demand): **$48,801,239**
Grocery sales estimate: **$21,463,750**
Grocery sales gap: **($27,337,489)**
Capture rate (sales/spending) = **44%**

Sources: American Community Survey (2008–12 5-year estimate);
Consumer Expenditure Survey (2011); US Census Zip Code
Business Patterns; Urban Land Institute (2008).

SUMMARY

There is a wealth of free, accessible public data available to build a descriptive and comparative profile for urban revitalization in a variety of contexts. This chapter has provided a detailed overview of how to define a community trade area and the important sources of public data available to understand a community, its people, and its surroundings. The chapter also highlighted the use of community indicators and basic techniques to better understand the social characteristics and market potential of a community. These indicators and techniques allow us to build a community profile that includes descriptive and comparative information on the residents, customers, housing, workforce, and businesses in a targeted area, and on how it is changing over time. They allow us to monitor change, establish revitalization planning goals, and evaluate progress toward achieving those goals. Applied alongside the techniques covered in chapter 13, these methods help to establish a strong understanding of a community's characteristics and routes toward positive change.

STUDY QUESTIONS

1   What factors influence the size and shape of a trade area? How would you go about determining the trade area for your neighborhood?
2   The data discussed in this chapter allow us to build a wide range of indicators to monitor neighborhood change. If you were tasked to ensure that an adequate and affordable supply of housing exists in a given neighborhood, what data would you want to collect? How would you build a set of indicators to establish baseline conditions and evaluate change over time?
3   Take a look at the business inventory in Table 12.3. What are the community trade area's key strengths and weaknesses? How does the trade area differ from the comparison areas? How does this information help direct strategies for economic revitalization?
4   Based on the gap analysis in Box 12.2, would you argue that the community can support additional grocery stores? Why? What additional information would you need to make a determination on the market capacity for additional store space?

REFERENCES

El Nasser, H, and Overberg, P. 2012. "Census continues to undercount blacks, Hispanics and kids." *USA Today* May 23. http://usatoday30.usatoday.com/news/nation/story/2012-05-22/census-hispanic-black/55140150/1

Gibbs, RJ. 2007. "A primer on retail types and urban centers." *Better Cities & Towns* September 1. http://bettercities.net/article/primer-retail-types-and-urban-centers

Jacobus, R and Hickey, M. 2006. *Commercial revitalization planning guide*. New York: Local Initiatives Support Corporation.

Jennings, J. 2012. "Measuring neighborhood distress: a tool for place-based urban revitalization strategies." *Community Development* 43(4): 464–475.

Meltzer, R and Schuetz, J. 2011. "Bodegas or bagel shops? Neighborhood differences in retail and household services." *Economic Development Quarterly* 26(1): 73–94.

NNIP. n.d. "NNIP concept." National Neighborhood Indicators Partnership. www.neighborhoodindicators.org/about-nnip/nnip-concept

Smith, RE. 2011. "How to evaluate Choice and Promise neighborhoods." In *Low-income Working Families Project*. Washington, DC: Urban Institute.

Trade Dimensions. n.d. *Retail tenant directory*. Accessed January 20, 2014. www.rctailtenants.com

Urban Land Institute. 2008. *Dollars & cents of shopping centers*. Washington, DC: Urban Land Institute.

US Bureau of Labor Statistics. 2011. "May 2011 National industry-specific occupational employment and wage estimates: NAICS 451200 – Book, periodical, and music stores. Occupational Employment Statistics." Accessed 15 January, 2014. www.bls.gov/oes/2011/may/naics4_451200.htm#27-0000

US Bureau of Labor Statistics. 2014. "Quarterly Census of Employment and Wages." Accessed January 15, 2014. www.bls.gov/cew

US Census Bureau. 2014a. "Nonemployer Statistics." Accessed January 14, 2014. www.census.gov/econ/nonemployer

US Census Bureau. 2014b. "Quarterly Workforce Indicators. Longitudinal employer–household dynamics." Accessed January 14, 2014. http://lehd.ces.census.gov/data

US Census Bureau. 2014c. "Survey of Business Owners." Accessed January 14, 2014. www.census.gov/econ/sbo

US Census Bureau. n.d.-a. "American Community Survey: Handbooks for Data Users." Accessed January 14, 2014. www.census.gov/programs-surveys/acs/guidance/handbooks.html

US Census Bureau. n.d.-b. "Economic Census." Accessed January 14, 2014. www.census.gov/econ/census

Williams, S, Galster, G, and Verma, N. 2013. "Home foreclosures as early warning indicator of neighborhood decline." *Journal of the American Planning Association* 79(3): 201–210. doi: 10.1080/01944363.2013.888306

13
Field methods

INTRODUCTION

An important way to understand a community or a place is to observe it and talk to the people who live in, use, and know the area. Field methods are techniques that help both planning professionals and community organizers learn how people use or perceive a given place, and its meaning to diverse community members. They also help us gain a better grasp of what residents or other users value and how an area functions or is managed. The field methods that we discuss in this chapter complement the community assessment techniques that evaluate the economic, physical, and social characteristics and the area's market potential, covered in chapter 12. Field methods offer a qualitative dimension that allows us to uncover information and meaning that we simply cannot acquire from census and economic data alone.

Demographic information can show a community's potential as well as differences among residents that will be relevant when establishing a revitalization agenda. Age and ethnic diversity, for example, might suggest that there will be different priorities and perspectives in the community. Income, poverty, unemployment figures, or health outcomes can help to identify situations of concern to community members. This information does not explain community members' perspectives, values, or priorities, nor does it tell us directly about local cultures or past efforts to effect change in a neighborhood. Finally, it does not explain the causes of the situation or barriers to improving opportunities or outcomes for a given community. To understand these dimensions, we must turn to community-based knowledge, techniques or practices used successfully in other places, and the current state of knowledge about revitalization. Field methods are essential if we want to know first-hand what community members consider to be major issues. They can help us to see how a brownfield site, new retail development, gentrification, or any number of issues impact a community.

Three important sources of information help explain why communities change and the potential impacts of proposed revitalization schemes. First, revitalization professionals and organizers can draw on existing research about important community issues such as housing, gentrification, or segregation, to give only a few examples, to understand broader trends or causal relationships. Second, they can use field methods to understand community members' perspectives and learn how particular spaces function. Third, they can engage with the community to set priorities and receive feedback on proposed initiatives. We will discuss the first two sources of information in this chapter. In chapter 14, we discuss community participation in revitalization processes.

RESEARCH METHODS AND FIELD METHODS

As we begin, you might ask what field methods are and how they differ from research. *Research* can be defined as a systematic investigation into a subject to develop knowledge and understanding. Research can be a slow process. Some research projects take years to complete and they build incrementally on prior research to contribute to collective knowledge about a subject. This is a valuable process. Without incremental knowledge building, we would know less about the effects of pollution, how the economy became globally connected, or the factors that lead to the gentrification of urban neighborhoods. Revitalization initiatives, in contrast, must gather information quickly and understand local dynamics as the basis for informed action. Field methods draw on research techniques but they have been adapted to be effective in planning and revitalization efforts. *Field methods* therefore can be defined as a systematic inquiry in a community or place with the intent to collect information to use as the basis for planning or revitalization initiatives.

Before we discuss field methods, we briefly discuss why understanding existing research can be a useful part of a revitalization initiative. Too often, research is seen as abstract or too far from everyday reality to be useful in problem-solving. This is not the case. For example, researchers have asked all the following questions. How effective are different policies intended to reduce vacant land? In what circumstances does tax increment financing work well? How do former residents fare after public housing is redeveloped? What policies and other factors cause observed patterns of disinvestment and reinvestment? Existing research therefore is useful for understanding a *trend*, a *phenomenon*, or the *impacts* of a tool or policy. To be relevant, research does not have to be conducted locally. Instead, a deeper understanding comes from looking across different cities and situations.

A single study does not create the basis for action. Rather, a synthesis of the research bridges the gap between knowledge developed by long-term research projects and effective local action (Campbell 2012). Think tanks, university research institutes, and professional organizations produce publications that synthesize the current state of knowledge in the form of reports or best practices. The Brookings Institution, RAND Corporation, Policy Link, and the American Planning Association are some of the many national and international think tanks that offer research analyses relevant for urban issues. Conscientious professionals and community organizers will take advantage of relevant information at the outset of any project.

UNDERSTANDING COMMUNITY DIVERSITY

Another initial consideration when approaching a revitalization project is local diversity. Urban, suburban, and—increasingly—rural areas are diverse. On the surface, diversity implies that people are different and those differences can shape what they want or need in local environments. The realities are more complex. Differences are based on cultural practices, ability, and life stage. They also reflect both historic and contemporary injustice as well as deeply entrenched inequalities. These factors are correlated with property ownership, income, and levels of disinvestment. As a result, most projects will impact people in various ways and diverse residents will respond differently to proposals. They will also respond in various ways to attempts to elicit their perspective about proposed changes. A demographic snapshot that describes the social and neighborhood characteristics helps to show who lives in an area and is a starting point to understand who will be impacted. This is important when designing field method protocols and participatory processes. Even if the project instigators come from the community, they likely understand some perspectives more than others. They might know that a situation is occurring but not necessarily how diverse participants experience it. Local residents also might be more empathetic to some neighbors than others.

Basic demographic information can contextualize who is present and absent at a meeting, or in a park or another urban space. Race, ethnicity, immigration status, languages spoken at home, age, household or family situation, and housing tenure are all factors that have been shown to shape people's perspective and priorities as well as influencing responses to proposed changes. Instead of assuming a particular perspective, however, it is important to seek information from diverse people and particularly from those you anticipate will have different perspectives.

FIELD METHODS

Once the project team understand who lives in an area, they can employ field methods in order to better understand a particular place, neighborhood, or community. Field methods help revitalization professionals get to know the communities they work with and the areas that will be impacted. In this section, we explain the most common field methods and what they do well. Think of these as a toolkit. You will not use each one in every situation. Because data collection and analysis takes time and resources, it is important to decide what type of information will be most useful and in what form. This will help determine which methods to use.

As we stated earlier, field methods are a systematic inquiry in a community or place with the purpose of collecting information to use as the basis for planning or revitalization initiatives. In this section, we explain the most common field methods used in urban planning and revitalization and the general guidelines that pertain to these methods.

Field methods can provide both quantitative and qualitative evidence. *Quantitative* data refer to those that can be conveyed numerically, for example, what percentage of a population prefers option A over option B. *Qualitative* data refer to those that are conveyed textually or in narrative form, such as "residents repeatedly discussed how they had to wait to cross the street and then rush to avoid fast traffic because there were no marked crossings; they also noted the children would often cross midblock surrounding the park."

Quantitative information is better at conveying the extent or intensity of a problem (40% of lots are vacant), the relative importance of an issue (80% of all residents surveyed considered abandoned property the most important problem), or the intensity of use (400 children played at the playground during a four-hour period). The Census and Bureau of Labor Statistics data discussed in chapter 12 are examples of quantitative data.

Qualitative information is better at explaining how a situation plays out ("children play on the streets in this neighborhood; on one street, they have fashioned a swing from a tree and on another they have set up a basketball net") or what residents think or say ("residents say the city has promised to reduce blight for years and nothing has happened"). A combination of qualitative data from the following field methods and quantitative data explained in chapter 12 and the survey methods described below will provide a detailed understanding of the neighborhood or community.

GUIDELINES FOR FIELD METHODS

Regardless of the methods you use, it is helpful to remember the following guidelines.

- Before commencing research, set out clearly what you need to know, and how best to obtain that information. Otherwise it is possible to obtain more information than you can use, which can result in a loss of resources and time.
- Develop a protocol that explains the observation plans or interview and ethnographic techniques. A protocol is a set of step-by-step instructions for conducting the field method effectively. The protocol should be systematic and intend to maximize variation (see the following two points). It should include techniques for recording the information. A clear protocol can also help ensure that different people collecting information will return with comparable information.

■ Be as systematic as possible. If you want to understand what neighborhood residents think about a proposed park redesign, it is more systematic to contact every third house, or two houses on each block, than to talk to people in the park, because there might be important differences between park-users and nonpark-users. If you want to know how a park functions, in order to capture the range of park activities, it is better to observe the park at different times of day and days of the week to, than at the same days and times.

■ Maximize variation rather than attempting to be comprehensive. You will not be able to talk with everyone who is impacted by a proposal. Use the demographic information that you have available to identify diverse stakeholders and attempt to ensure you collect information across relevant differences among people. If necessary, use multiple techniques to gather information from different people. People with different incomes, or of different ethnicities, ages, and household situations (children living at home, for example), might have different but relevant perspectives and concerns.

■ Be flexible. Even with well-laid plans, unexpected contingencies can arise. As you begin to talk and observe, your perspective on what you need to know and what is occurring might change. You should be prepared to adapt to unexpected circumstances and information.

■ Aim questions and observation directly at your target. This might seem obvious, but it is surprising how often people disregard this basic point. If you want to know what seniors in a community think, ask the seniors rather than listening to others speak on behalf of seniors. If you need to know who uses an area slated for redevelopment, observe the area. You will obtain more accurate information about how a park is used by observing the park than by only asking people how they use it.

■ Finally, use the data that you collect. Collecting data is only the first step toward revitalization planning. Devise a plan to analyze the information you gather and to present it in useful ways. Writing a straightforward report will ensure the information is available later for others who are interested in the area or the project. You might need the information for an official document, grant proposal, or decision-making process. It can also be useful in public meetings.

DIRECT OBSERVATION

Direct observation describes a systematic approach to watching what occurs in a given area to explain its characteristics and how it functions. No other form of information can replace what you learn from going to an area and carefully looking around. The data and techniques we describe in chapter 12 are critical to understand the local area and the local market, but they cannot capture the intangible elements of being somewhere: the neighborhood's character; the specific site characteristics of a given lot; whether vacant properties are well maintained or filled with trash; or when the streets and parks are active or quiet. When revitalization professionals work in an area where they do not live, spending time there also gives a basis for understanding and communicating with residents about issues of importance to them.

Observation refers to methods that primarily involve watching and recording people's activities and the relationships they have with the built environment. In many cases, in addition to getting a general feel for the place, revitalization professionals or local organizers want specific information about particular places. How well used is a park, and who uses it? How do conflicts between cars and pedestrians in a shopping district play out? Do food trucks cause sidewalk congestion? These questions can best be answered by direct observation.

Using *field notes*, or the notes taken during and after observation sessions, the observers describe what has occurred as well as the weather, the tone, differences among people present, and other relevant qualitative factors. It is critical to take field notes if the information is to be used later. Unusual occurrences are often best captured in notes.

It is also useful to map where activities occur. *Behavioral mapping* is a recording tool to show how a space is used (Low, Taplin, and Scheld 2005). On a behavioral map, observers record what activities occur where. Identifying different types of activities and then counting their frequency as well as who engages in them (seniors or children, men or women, and so on) can convey intensity of different types of use. Detailed field notes are used in conjunction with behavioral maps to record in detail how activities—or conflicts—play out.

Observations also include *mapping physical traces* of activities (Zeisel 1981). When surveying vacant property, you might note informal activities that are occurring. In areas of high vacancy, many neighborhood residents maintain and use abandoned properties and therefore have reason to want access to that property. Obvious maintenance, play equipment, and gardening can show a lot is being used. Other physical traces in a neighborhood might include, but are not limited to, discarded needles or alcohol bottles, signs of sleeping or living, indicators of animals or people caring for feral animals, and informal pathways indicated by erosion or trampled plants. These show patterns of use that might be less obvious during observation hours.

Direct observation can be time-consuming. To understand a place, even a single street, observations should occur at different times of day and on different days of the week. This helps to ensure the observer captures the spectrum of activity in a neighborhood and does not misrepresent atypical events. Likewise, property condition surveys need to occur block-by-block because conditions will vary over the neighborhood.

Most people involved in revitalization efforts will likely engage in *participant observation*. Community-based revitalizers and outside professionals alike become participant observers as they attend and organize meetings and events, use the parks and streets, and participate in myriad ways. Participants learn much about an area. Participant observation over a short period, however, is rarely systematic. Even a neighborhood resident, for instance, understands local activities when she walks with her dog in the evenings or takes children to the park on weekends, but she might know less about an area during the hours when she is at work. Participant observation, nevertheless, can be important, and the informal discussions that occur are invaluable to understanding an area or a community's priorities or concerns.

SEMI-STRUCTURED INTERVIEWS

The best way to learn what people think is to ask them. This common-sense idea is the basis of conducting *interviews* or asking a series of questions about a topic of interest. Interviews can provide rich information about what residents think about their neighborhood, what they want to happen, and what has been done in the past. To conduct interviews systematically, the revitalization organizers must decide *who* to interview, *what* to ask, and *how* to record the information.

Stakeholder interviews, or interviews with people who will be impacted by an urban revitalization plan or project, seek to understand varied perspectives on a project or a topic. When individuals are contacted because of their status as community members (as residents or local workers, or people who use a particular facility, for example), the interviewers seek to speak with a *sample* of stakeholders. Because qualitative interviews rarely obtain probability samples (discussed later in the survey subsection), the interviewers seek to be as systematic as possible and maximize the range of differences among subjects to ensure the interviews are as representative as possible.

Systematic interviewing has challenges absent with direct observation. The interviewer has to make contact with participants and they must agree to be interviewed. Given the mix of cell phones and land lines, and the frequency with which people move, there is no reliable way to obtain all phone numbers of residents in a neighborhood. As a result, walking door-to-door is often a more reliable way to reach residents systematically. As the interviewer or interviewers walk door-to-door, successful interviewing depends on: (1) whether someone is home; (2) whether the person who is home is the right person with whom to speak; and (3) whether he or she has the time or willingness to speak at the time you arrive. It will take multiple visits to an area to obtain necessary interviews. Unless the area is very small, it will be impossible to talk with everyone, and choosing every third house or a reasonable subset of houses will be more systematic than talking to the first ten people who will talk. It is also important to ensure the residents interviewed reflect the area's diversity. Collecting basic demographic information can help assess how well the participants reflect the community.

Because systematic interviewing is time-consuming, it is more common to conduct focus groups or a survey, or to rely on public meetings and participatory processes to hear residents' perspectives. Nevertheless, in some cases the local dynamics might make the intensive engagement in interviewing worthwhile. For example, if the community is diverse and recently changing, unknown perspectives might make the front-end work important to reduce unanticipated responses to future proposals. If particular subsections of the populations will not participate in public meetings or focus groups, interviews can fill in gaps.

Two other interview approaches can be useful. *Expert interviews* refer to contacting people based on their position at a particular institution or in the community. For example, a church pastor, an independent grocery store owner, or the manager of a regional park might be contacted because they have special knowledge based on their position. In any process, it is often useful to reach out to community leaders for an informal or formal interview. Interviewers can draw on a *panel of informants* if they seek to understand the impact of a situation, place, or event on a community. In this case, they interview the range of people who are familiar with the situation so that collectively the interviews give a reasonably complete picture (Weiss 1994). For example, in order to determine how successful a local festival was at attracting community members and showcasing the neighborhood, the interviewers might interview a sample of people attending the festival, the festival organizers, performers, vendors, and local businesses.

Most interviews used in planning and revitalization take the form of *semi-structured interviews* where the interviewer asks specific but open-ended questions on directed topics. When conducting interviews, it is important to develop clear questions that everyone will understand. Testing questions prior to beginning interviews is a necessary first step because it often takes a few tries to ask a question in a way that elicits the information the interviewer seeks. It is equally important to be prepared to record people's answers. If the interviewer intends to take notes, be sure to build in enough time directly after each interview to flesh out the notes that are jotted down during the interview. While interviewing, the interviewer must keep listening and not become absorbed in note-taking. Interviews also can be recorded with inexpensive digital recorders or mobile phones. In this case, resources to transcribe the interviews are necessary. Some people will decline to be recorded, and more people will decline to be videotaped if asked, so note-taking should always be available as a backup.

A technique that combines elements of an interview with direct observation is a *transect walk* with a community member. A guided walk through the neighborhood can be an opportunity for the community member to describe and comment on the neighborhood. In this case, the interviewer does not primarily ask questions—although she or he can ask follow-up questions—but instead listens to what the resident says about the neighborhood and follows what the resident wants to show. This can help develop a community-centered understanding of the area (Low, Taplin, and Scheld 2005). Community activists can often provide invaluable, in-depth knowledge of an area's history and previous revitalization efforts.

FOCUS GROUPS

A *focus group* is a small group discussion with a select group of participants. Similar to semi-structured interviews, the questions are focused but open ended. The focus group format facilitates participants' discussion about a situation or local conditions which can help participants to articulate shared concerns or neighborhood issues or show different perspectives on the issue. This can be an advantage over individual interviews in which people do not hear other perspectives when they answer questions.

Because focus groups are intended to be in-depth discussions, groups should be limited to no more than six to eight people to ensure an inclusive discussion can develop. A skilled facilitator is necessary to bring everyone into the discussion and allow flexibility as important issues arise, while staying focused on the group's purpose. The topic can be a specific project but it can also be about problems that have occurred, general perceptions about an area or neighborhood, or community priorities. Multiple focus groups can be held to reach more people or more groups of people.

Focus groups can target a cross-section of the community, or different focus groups can be held with different groups within a community. Because focus groups can be particularly useful to obtain the perspectives of a particular segment of the population, such as people with disabilities, youth, or recent immigrants, it is more common to hold focus groups with people who share relevant characteristics. What characteristics are relevant will differ by situation, and the organizer must take time to understand local dynamics when determining who to invite and how to contact participants. It is also important that people can contribute in their primary language or one with which they are comfortable.

Focus groups have disadvantages when compared with interviews, however. They can cause people with perspectives outside the norm to be silent if they do not believe the perspective will be respected. Participants may also be hesitant to discuss sensitive or controversial topics.

Because focus groups can both show common perspectives and differences among perspectives, it is important to record and transcribe the discussion. Otherwise the nuances will be lost as the discussion develops. If this is impossible, one or two note-takers in addition to the facilitator may be necessary to record the conversation.

SURVEYS

Surveys are the most familiar type of field method. In everyday life, survey information is everywhere. Gallup public opinion surveys tell us what people think. Consumer preference surveys tell us which products people prefer. Consumer expenditure surveys show us how people spend their money. The US Census captures who lives where and various characteristics about the residents. *Surveys* are a formal method that obtain relatively large numbers of responses on specific questions. According to Singleton and Straits (2005: 239), professional surveys have three common features:

- a large number of respondents are chosen through probability sampling procedures to represent the population of interest
- systematic questionnaire or interview procedures are used to ask prescribed questions of respondents and record their answers
- answers are numerically coded and analyzed with the aid of statistical software.

Probability samples are samples collected using a random sampling technique that is designed so that it is very likely that the sample reflects the population from which the sample is drawn or, in other words, that the sample can be generalized to the population of interest. In revitalization processes, surveyors often cannot collect probability samples and instead must rely on systematic and representative samples, similar to interviews.

Although surveys and interviews are similar in that they ask people questions about what they think, they have different strengths. Surveys create data that are uniform, and the breadth of contacts allows for more certainty that the findings reflect the perspectives of the people surveyed. Interviews offer information with more depth and nuance, but they sacrifice uniformity (Singleton and Straits 2005, Weiss 1994). Survey questions are designed to be easily quantified. For example, a survey could ask residents to choose among different possible local amenities and quickly determine the four major priorities and their relative importance.

Surveys are administered either by interviewing respondents or by using a written questionnaire. Over a small area, written questionnaires can be sent to all households, making it easy and affordable to be systematic. It is likely that repeated attempts will be necessary to elicit responses from in-person or mail surveys. Surveys can also be administered to people who use an area by approaching them and conducting the survey on site. Various free or inexpensive online survey tools are available. These are very inexpensive and allow for widespread participation, but the responses will not be systematic. As with interviews and focus groups, the surveys should be offered in the language with which the respondent is most comfortable. When considering written surveys, local literacy rates should also be taken into consideration.

A wide range of additional survey types can be conducted during revitalization initiatives. These include property conditions and vacant land surveys, business surveys, street tree surveys, and surveys of the condition of streets, parks, or other infrastructure. In this case, whenever possible, the intent would be to survey the entire area of interest.

HISTORICAL AND ARCHIVAL DOCUMENTATION

In some situations, evaluation of historic and archival information will be useful. Historic research is necessary for sites that might be designated as historic landmarks. It is also a necessary step to identify uses that might have contaminated a site and resulted in a brownfield. A historical analysis also can show how a neighborhood has changed and bring forward relevant information that makes contemporary conflicts or sources of tension clear. The public library, local historical societies, university special collections, and city and county archives all might have archival sources relevant to a given site or area. Sanborn insurance maps are detailed maps of cities and towns that show land use on each parcel. They date back to the nineteenth century, and many cities have historical maps available that can show how a neighborhood or block changed. Property records, which are available from the county tax assessor, can also be useful to identify the history of uses on a site. Census data from multiple decades can also show how an area has changed. Lifelong residents can be another source of invaluable information. Understanding this change can give context that will help participants understand the concerns raised by current residents.

RAPID ETHNOGRAPHIC ASSESSMENT PROCEDURES

When attempting to understand cultural differences and values, rapid ethnographic assessment procedures (REAPs) can be useful. *REAPs* draw on various methods to understand how people act, what they value and want, as well as the varying perspectives of different stakeholders, from ordinary citizens to managers and professionals. The purposes of these methods are to be holistic and understand both use and values from different perspectives. They are based on ethnographic methods which were developed to understand perspectives and situations from the participants' points of view.

Rapid ethnographic assessment procedures may include many of the sources and methods we have already discussed, including historical and archival documents, mapping physical traces, behavioral mapping, transect walks, individual interviews, expert interviews, impromptu group interviews, focus groups, observation, and participant observation. REAPs are intensive methods that require significant time investment, but can also give robust understanding of an area. Because these methods occur over short periods of active effort, community consultants and volunteers can be trained in data collection.

Using this method to analyze large urban parks, Low et al. (2005) spoke with park-users across different parts of the parks during different times of day, as well as park managers and other official sources. They also analyzed the built environment and observed park use as well as how users adapted the spaces. Through this process, they learned how different stakeholders used the park, what they valued in the park, and what were opportunities and limitations from the park managers' perspectives. They showed differences among park-users including but not limited to seniors, families with kids, large groups, gay men, and different ethnic groups. These processes also uncovered sources of conflict. For example, group picnickers wanted grassy shaded areas for large group picnics and, as a result, disregarded park managers' restrictions on sitting under trees in grassy areas that were intended to be restored. While the restoration was critical to the park managers, it was less important to the picnickers.

When one uses REAPs, it is best to be flexible. Observing a popular area might lead to a conversation with teenagers using the area. A request for information might result in a transect walk if someone does not want to stop, but is willing to let you come along. Taking advantage of situations as they arise is an important dimension in REAPs (Low et al. 2005).

FINAL THOUGHTS ON COLLECTING FIELD DATA

The field methods that we describe provide different types of information. No form of data is more "true" or accurate than another. Quantitative and qualitative evidence both offer useful information. We collect various types of information because revitalization processes are complex and involve many different stakeholders who have different ways of knowing and making decisions.

Keep in mind that field methods can be difficult. Do not be surprised if people do not want to talk with you, or act skeptical when you approach them. Remember, your objective is to develop the best possible revitalization projects and to do that you need good information about communities and local users. Be adaptable in the field and ask repeatedly who has contributed to the discussions and who has been absent.

ANALYZING, REPORTING AND USING INFORMATION

Once you have collected your data, you must analyze and report it in a useable format. This section explains analysis and reporting techniques. The purpose of collecting the data is to provide good information as the basis for action. The data can be used in different ways. They can serve as an important basis for proposed projects or a discussion with developers about possible amenities in a project area. They can be presented to community members in forms that will be useful for community initiatives and grant proposals. They can also start a discussion at a community meeting.

Before you present the results of your data collection efforts, you must analyze and interpret your findings. *Analyzing evidence* refers to synthesizing and understanding the data. *Interpreting evidence* refers to explaining what it means and determining its relevance in the context in which you will be using it. Two common analysis techniques are coding and triangulation.

Coding refers to the process where major categories or themes are developed. If you are conducting interviews and neighborhood residents have mentioned the need for jobs, a park, or addressing blighted property, these would become themes or categories. Sometimes people talk indirectly about an issue and you will only understand the meaning once you better understand the situation. For example, when someone mentions that young people have too much time during the summers, they might be discussing too few job or recreational opportunities. Once you have developed a set of themes, you will identify each time or instance the themes come up in the evidence. For example, if crime or violence is an important theme, the "instances" in the evidence might take different forms. They might include physical remnants of drug use, bars on windows or other indicators of fear or security, as well as residents' comments on crime or drug dealers. However, if teenagers hanging out on the corner are identified as the problem, and you speak with the teenagers, their perspectives would also be included even if they are counter-evidence—perhaps they state that they just spend time there to talk. When you analyze a theme that you identified, you must incorporate all perspectives on that topic. A given piece of evidence or instance may fall within more than one category, such as both crime and unemployment.

All qualitative evidence, from field notes to interview transcripts, is coded. Coding can be as straightforward as reading transcripts and field notes and highlighting different themes and the different instances. Quantitative evidence can be entered into a spreadsheet or database in order to analyze it. Keep in mind that counting instances in qualitative evidence can also be insightful. Qualitative analysis software is available that can help organize the evidence if you anticipate doing a significant amount of work over long periods of time.

Triangulation refers to the process where different sources of evidence are used in conjunction with one another to obtain a more complete and accurate picture. Consumer expenditure data can show how much money a community spends. A market analysis can give an approximation of how much of that is spent locally. Focus groups can help explain where people shop and their travel patterns when shopping. In the case of blighted property, multiple sources of data can help give a more complete picture of the issue: a property survey can help identify which lots are abandoned, residents can explain their responses to blight in interviews, and city records can show ownership information and actions the city has taken. Together, the different sources of information allow you to consider blighted property from multiple angles and, quite possibly, identify more appropriate solutions.

Interpretation is the process of determining the most important themes and how to convey and report common and less common perspectives or, in other words, how to explain what is going on. As you decide how to present the data and write a report, you are essentially interpreting the information you have found. You might get it wrong or partially wrong. However, making the information available to participants and listening to their responses can help make the final report more accurate. Continue to be adaptable. If, after much discussion, you realize an interpretation is not accurate, change it. The objective is to develop more useful information rather than defending a particular perspective.

The final step is to present the data in ways that are available and useful to interested stakeholders. If you have gone through the effort of collecting evidence, put it into a form that others can draw on in the future. A straightforward report (with images and graphics) might be the easiest way. Be sure to describe the context and methods or they will be lost.

EVALUATION RESEARCH

Evaluation research is another type of research used in revitalization initiatives. Because it has a different purpose, we discuss it separately from field methods. Evaluation research can be defined as

> the use of social research procedures to systematically investigate the effectiveness of social intervention programs that is adapted to their political and organizational environments and designed to inform social action in ways that improve social conditions.
>
> (Rossi, Freeman, and Lipsey 1999: 20)

Since the Great Society Programs and the War on Poverty, policymakers have asked for evidence that a policy or funded program will have the desired effect. Increasingly, granting organizations and public agencies want to see measurable impact from their investments too. As a result, evaluation is increasingly built into project proposals in order to determine whether the project or program is meeting its objectives.

Similar to other field methods, evaluation research is adapted to provide specific information that is useful to an organization, an organization's clients, granting agencies, and at times other people who might be considering a similar initiative. In some cases, evaluation can show weaknesses that can be rectified and immediately improve how a program or area functions. Evaluation research can be performed by an outside evaluation firm or other experienced evaluation researchers. Because resources are often a limiting factor, organizations might opt to conduct their own evaluation. If the evaluation will be done internally, it is worth seeking some additional references to help guide the effort.

As with other field methods, a first step is to determine what information is desired. Evaluations may have many objectives, such as determining project costs per person served, or if the program is achieving a defined purpose such as workforce readiness, job creation, or vacant land reduction. A comprehensive evaluation would include an assessment of the need for the program or project, its design, implementation, impact, and efficiency (Rossi et al. 1999). The evaluation sponsor, however, can decide which factors are most important and, in many cases, only particular criteria are evaluated.

Evaluators must develop a description of the program or project performance that will be evaluated and a series of criteria that will be the basis on which it is judged. The evaluators then determine the best way to answer the questions, given the constraints imposed by the situation or by the available time and resources. Evaluation may involve any of the methods discussed above (or additional methods). It is important to keep in mind that the impact of a project or program comes after it is operating or substantially complete, but some evaluation procedures benefit from collecting data prior to the intervention or project completion to establish a baseline against which to gauge future change. If noise or air pollution from a project is a concern, for example, collecting baseline data before the project is put into service is necessary. Early attention to evaluation will save frustration later after opportunities for data collection have passed.

SUMMARY

Field methods describe a range of ways to collect information about a community. They complement other assessment techniques and are uniquely able to provide qualitative information about priorities and perspectives that are lacking from most national data sources. Field methods draw on research method techniques that promote systematic inquiry and reduce bias in the data sources that have been adapted for rapid use in urban planning and revitalization processes. The most important point to remember is that nothing replaces being somewhere, talking with people, and observing the situation. Local stakeholders bring this knowledge. While this does not offer complete information, it augments the other information that professionals regularly use in their working lives.

STUDY QUESTIONS

1   Explain the difference between qualitative and quantitative information. What can each answer?
2   What are the main challenges that arise when conducting field methods?
3   You are working with a coalition of residents and local businesses to improve local conditions. After speaking with them, you identify two potential issues that the coalition could tackle. You are tasked with writing a two- to three-page report on each topic to help determine which of the two would be the best place to start. Choose two issues and use three different techniques to identify national practices that appear promising for your situation: (1) use an article database available from the library; (2) search the web for research using Google Scholar and Google Books; and (3) use Google to search for professional reports and organizations. Briefly explain the differences in the results you find. Did your analysis help you determine the best course of action? Why, or why not?

4   Your coalition has chosen a series of neighborhood improvement actions. Describe what you want to accomplish. Next, describe three field methods you could use to collect local information to inform an effective plan. What will you learn from each? How will this information help you design more effective plans for action?

REFERENCES

Campbell, Heather. 2012. "Planning to change the world: between knowledge and action lies synthesis." *Journal of Planning Education and Research* 32(2): 135–146. doi: 10.1177/0739456x11436347

Lofland, John, David Snow, Leon Anderson, and Lyn H Lofland. 2003. *Analyzing social settings: a guide to qualitative observation and analysis*. Belmont, CA: Wadsworth.

Low, Setha, Dana Taplin, and Suzanne Scheld. 2005. *Rethinking urban parks: public space and cultural diversity*. Austin, TX: University of Texas Press.

Rossi, Peter H, Howard E Freeman, and Mark W Lipsey. 1999. *Evaluation: a systematic approach*, 6th Ed. Thousand Oaks, CA: Sage.

Singleton, Royce A and Bruce C Straits. 2005. *Approaches to social research*. New York: Oxford University Press.

Weiss, Robert Stuart. 1994. *Learning from strangers: the art and method of qualitative interview studies*. New York: Free Press.

Zeisel, John. 1981. *Inquiry by design: tools for environment–behavior research*. Monterey, CA: Brooks/Cole Publishing Company.

14

Public participation

INTRODUCTION

The previous two chapters have discussed different sources of data that are central to developing revitalization initiatives. Public participation is another source of local knowledge as well as a critical process for ordinary residents to engage with proposed projects and discuss shared concerns. Public participation can be considered democracy in practice because it provides regular people with the opportunity to be involved in day-to-day decisions that affect them (Beierle and Cayford 2002). As Diane Day (1997) has argued, however, public participation is "an essentially contested concept." Although it is good for its own sake and necessary in a democratic society, participation can be at odds with planning practice. Instead, planning is often driven by local politics, technical expertise, and professional knowledge (including those tools discussed throughout this book).

Nevertheless, it is the regular people who work, live, and spend time in an area who will be impacted by development policy and projects. Using public money for private development is justified on the grounds that it will provide community benefits, and most US cities and states require that residents be informed about new development in their neighborhood. Participation has become a buzzword used widely by people ranging from radical activists to local governments and global governance organizations. Public institutions are responding to the demands that ordinary citizens be involved in decision-making processes, but at times, observers fear that the outcome is manipulation or coercion rather than influence and partnership (Arnstein 1969, Cooke and Kothari 2001, Cornwall 2008).

The field methods we discussed in chapter 13 work in conjunction with participatory processes whereby residents and other ordinary citizens engage directly with a proposed project or policy by providing feedback and shaping outcomes. In this chapter we discuss the techniques and challenges with public participation. In next section, we define participation and outline common participatory mechanisms. The following section considers how to work with the challenges that arise. In the final section, we discuss action outside of formal participatory processes and how it shapes development outcomes.

WHAT IS PARTICIPATION?

In this section, we discuss the context in which participatory planning arose and what it intends to accomplish. We then briefly describe the most common participatory tools.

A BRIEF HISTORY OF PARTICIPATORY PROCESSES IN PLANNING AND DEVELOPMENT

The need for participatory urban redevelopment reaches back to the early twentieth century. Urban initiatives—often promoted by reform organizations—reflected a growing belief that public intervention could increase environmental quality and prosperity in urban areas. It ushered in a period of modernist urban planning when large-scale plans and projects took precedent over neighborhood concerns. Public housing construction in the 1930s was accompanied by "slum clearance" efforts that demolished neighborhoods. Mid-century freeway construction took out whatever neighborhoods stood in its path and was demolishing more than 62,000 housing units annually by the late 1960s, according to the US House Committee on Public Works (Mohl 2004). These and other development decisions destroyed many neighborhoods without discussion with existing residents. The actions destabilized housing and social networks and caused a sense of loss and anxiety. As a result of these demolitions, residents became worse off than when they were living in the "slums." Not all neighborhoods were valued equally, and those where people of color lived and with low-income residents were more frequently bulldozed (Gans 1982, Marris 1986, Mohl 2004, Vale 2013).

Ordinary residents demanded the opportunity to participate in and influence planning and redevelopment efforts that impacted them and their neighborhoods. Many planners and others who observed urban redevelopment at work agreed that different approaches were necessary. In the following decades, planners developed equity, advocacy, and communicative planning techniques to address the processes leading to disparate impacts, to give community members technical assistance and create ongoing engagement (Fainstein and Campbell 2012, Healey 2006).

Participation in the US was first formalized in the Economic Opportunity Act of 1964 (EOA) which, as we discuss in chapter 3, called for the creation of a diverse set of programs focused on neighborhood revitalization and alleviating poverty. One of the key components of the Act was the requirement that the programs be "developed, conducted and administered with the maximum feasible participation of residents of the areas and members of the groups served" (quoted in Rubin 1969: 15). This participatory idea had roots in both economic development programs in the global South and pressure from the Civil Rights Movement. Through participation, the intent was that people could influence programs that impacted them while developing greater civic or community capacity to enable further future action.

Participation requirements were nevertheless met with resistance. Some welfare agencies claimed that it was impossible for poor residents to participate despite their desire to do so. Even well-meaning agencies found it hard to plan with communities rather than plan for them, and many professionals were surprised when residents demanded to be included (Rubin 1969). What constituted "maximum feasible participation" was debated, and it was the divergent outcomes that led to Sherry Arnstein's (1969) influential ladder of citizen participation.

As Arnstein (1969: 216) noted, "The idea of citizen participation is a little like eating spinach; no one is against it in principle because it is good for you." However, she also recognized that all participatory processes are not the same. After observing numerous attempts at participation, Arnstein proposed a "ladder" that distinguished between eight types of participation. The bottom rungs of the ladder include manipulation and therapy, which she considers to be nonparticipation. At the top of the ladder are three representations of citizen power: partnerships, delegated power, and citizen control. In between, she placed placation, consultation, and informing, approaches she called tokenism. Arnstein's ladder differentiates between the power of citizens to shape an outcome and the empty ritual masquerading as decision-making. Arnstein's eight-rung ladder of citizen participation is still widely recognized as community organizers and professionals alike struggle to ensure that participatory processes lead to beneficial outcomes.

DEFINITIONS

Community members and professionals have defined participation in ways that reflect the need for meaningful engagement. Fischer (2000: 1) defines participation as "deliberation on issues affecting one's own life." Cornwall and Gaventa (2001: 32) define participation as the "forms of inclusion, consultation and/or mobilization designed to influence larger institutions or policies." Saxena (2011: 31) states that "participation should include notions of contribution, influencing, sharing or redistributing power and of control, resources, benefits, knowledge and skills to be gained through beneficiary involvement in decision-making." Drawing on these ideas, we define participation as regular people's engagement in design and decision-making processes in ways that influence the projects, policies, and larger institutions for community benefit, and through which ordinary citizens gain skills to participate in and influence planning and development processes.

We use the terms "regular people," "ordinary citizens," and "community members" interchangeably. These terms refer to people who will be impacted by a project in some way. They are therefore stakeholders, but those who have no formal role as professionals in the process, or in the business and development communities. In this case, "citizen" does not refer to a person's country of birth or naturalization, but to the inhabitants of a place who have both the rights and the duties to participate in decision-making that impacts their communities. Ordinary citizens are situated differently from one another. In some places, residents have direct connections to the business and development community and can exercise influence as friends, business partners, and neighbors. However, in most instances, residents have few connections and must rely on their claims as people who have the right to a voice because a proposed change will affect them, or because they want to effect change.

The word community is often used in planning and urban revitalization to describe neighborhood residents because they share the characteristic of living in an area, but the notion of community can obscure differences and even divisions among people. While neighbors may be bound by a shared interest in one place, they may have very different histories and opportunities. These differences influence how people perceive events and their expectations about what will happen, as well as their priorities and their preferences. Innumerable instances of neighborhood conflict, over anything from the expansion of a church to new retail development, illustrate this.

In this chapter, we primarily discuss participatory processes that are instigated by the public sector or development team, regardless of whether the development team is based in the community—in other words, the lead agency or organization. These processes can be considered top-down. Not all forms of participation come from those who initiate the project, program, or policy. People can organize around a project or set of concerns and demand inclusion. Therefore participation can be top-down or bottom-up (White 2011).

THE OBJECTIVES OF PARTICIPATION

The immediate objectives of participatory processes may be considered at two different levels. First, what are the intended outcomes for a given process? Second, what are the societal benefits to implementing participatory processes? In this subsection, we describe seven main objectives to participation that incorporate both levels.

Incorporating local knowledge into projects or policies

Improving the quality of projects or decisions is a primary objective of many participatory processes (Beierle and Cayford 2002). Because locals may have expertise about the social, natural, and built environments, this information improves the substantive quality of projects and decisions. Ordinary citizens' knowledge therefore results in better outcomes, including those that meet the agency's or developer's goals. Conversely, the failure to consider and incorporate local knowledge has repeatedly contributed to serious problems, ranging from the "spatial mismatch" of jobs and housing in urban renewal-era public housing to a lack of green space in many urban areas.

Incorporating local priorities or values into projects or policies

A second, related objective is to incorporate public values and local priorities into projects or policies (Beierle and Cayford 2002). In this case, the objective is to ensure that a project reflects community members' interests. Local priorities can be specific, as in the case of a park design where the designers provide facilities for locally preferred activities such as soccer or picnics. Priorities may also be broad, as when an initiative is developed to address community concerns such as opportunities for youth or vacant land. In many cases, communities develop interests through participatory processes and political activity. Through talking and working together, collective interests and concerns are both developed and articulated (Stone 1987: 15).

Educating and informing members of the public, public officials, professionals, and developers

Sharing information is an important element of participatory processes. Ordinary citizens need adequate information to shape plans or projects (Beierle and Cayford 2002). Public officials, professionals, and developers need to learn community concerns. If the purpose of education is to convince people that they must accept a project or change their priorities or preferences, education can quickly become what Arnstein (1969) recognized as manipulation. If the purpose of education is to create a base of knowledge on which to develop interests, shape feedback, and take action, it is a necessary part of a participatory processes. Participatory processes often successfully inform people about what is occurring (Beierle and Cayford 2002), but this alone is minimal participation.

Building trust in institutions, plans, and projects

Another objective of participatory processes to is to build trust—or achieve buy-in—on a project or plan and, in the process, the larger public institutions (Beierle and Cayford 2002). If planning processes and development projects are designed to share power and allow community influence, ordinary citizens will learn that they can work with public officials and private developers to come to mutually beneficial outcomes. For projects and plans that will develop over long periods, intensive engagement that increases community buy-in and shows that public agencies are responsive and accountable can reduce later conflicts or opposition.

Resolving conflicts among different interests

In virtually all cases, different people have varying perceptions and priorities. Community members living near a project, whether it is a new school or a large retail complex, might oppose it because of concerns about the immediate effects of traffic and noise. People also oppose projects such as public housing or homeless shelters out of fear and bigotry. A participatory processes can help articulate the concerns and begin dialog to develop better understanding of a project and build mutually agreeable alternatives (Beierle and Cayford 2002). At times, resolving conflicts will require intensive negotiating and mediation, whereas on other occasions simple modifications can reduce potential problems.

Monitoring or stopping potentially destructive actions

In some cases, local growth coalitions take action that will have adverse impacts, and community members organize to stop a proposed project (Mohl 2004, Stone 1989). Local residents might learn about a project through formal information channels, or they might hear informally from neighbors that a valued site may be harmed. Community-based organizing has successfully stopped projects, helped to mitigate adverse effects, and gained concessions. For example, community groups may reach agreements with local government to replace destroyed housing with new units or acquire new parks or playgrounds in compensation for unwanted development.

Building participatory democracy

Engaging and organizing can contribute to ordinary citizens developing the capacities to effect change. Engaged people can share power and better influence their local environments. Participants develop what Hanson (2003) calls "civic capital," or the civic skills to be active participants in local issues, by engaging in processes to effect change. For civic capital to be produced and deployed, officials and citizens alike must "receive positive feedback from their participation in the form of policy successes, personal satisfaction and the development of civic skills that can be applied to other issues" (Hanson 2003: 323).

In New Orleans after Hurricane Katrina, for example, activist Karen Gadbois began a blog called "Squandered Heritage" to raise awareness about demolitions of historic buildings that could provide housing. After observing city processes and working to stop unnecessary demolitions, she realized that there was a great need for in-depth news coverage of redevelopment, environmental issues, and other areas of local concern. She used the expertise she had developed and became a founding member of the city's first nonprofit newsroom, *The Lens*, which offers in-depth reporting on regional issues (thelensnola.org).

While working on civic issues, people learn to work in their own and their group's interest at the same time as working for collective interests across race and ethnicity, neighborhood, and socioeconomic class. Part of this activity involves building and maintaining alliances and coalitions where diverse views are represented (Hanson 2003).

WHO, WHEN, HOW AND FOR WHAT?

When designing the participatory process, numerous decisions will need to be made. The critical questions can be briefly summarized as Who? When? How? and For what purpose? (Cohen and Uphoff 2011, Saxena 2011). In addition, organizers must keep in mind why residents take time to participate.

Who will participate?

A primary step in designing the participatory process is to identify potentially impacted stakeholders. Impacted community members can be thought of geographically and as communities of interest. Basic demographic information can show the diversity in a geographic area and can be useful to determine who needs to be brought into the discussion. Demographic information can also help determine what type of outreach will be necessary, which languages are spoken, and other factors such as family structure might influence the participatory process design. It can also help assess whether some stakeholder groups are underrepresented during the process. Communities of interest might include housing, civil rights, or environmental organizations that might be citywide or community-based.

When will people participate?

Another decision will be to determine when regular people will be brought into the process. Building knowledge about scientific and technical issues takes time. Residents might need information in time to talk with others and understand the implications of the new proposals. For complex redevelopment initiatives, different types of outreach will be necessary at different times. It is worth remembering that participation cannot be turned on and off. People may start organizing against a project if they hear of it and are not included, and they might continue taking action after a formal participation process ends. Reaching out to community members early can start a dialogue about concerns as well as desired outcomes for the area.

How will people participate?

This question asks which mechanisms will allow sustained engagement. The most common participatory mechanisms—public meetings, advisory committees, and negotiations—are discussed in the next subsection. No tool is inherently good or bad. An advisory board that intends to educate people or engineer their support would only be manipulation in Arnstein's (1969) ladder. A public meeting that allows people to state their frustrations but makes no room for change could be considered "therapy." Conversely, an advisory board with authority can create knowledgeable community members who can shape and even champion a project. A public meeting at the right moment also can provide information to many people simultaneously.

For what purpose will people participate?

In other words, what are the intended outcomes of the participatory process? The process needs to be designed to create the conditions for those outcomes to be realized. A process to work through anticipated differences would be designed differently from a process to inform people about future investment or to obtain feedback on previously defined priorities. To be effective, the purposes must be identified so the process can ensure the objectives will be met.

Why will people participate?

Ordinary citizens participate by initiating action or responding to a formal process when they see it as in their interest. Their interests can change. Residents might arrive as collaborators and leave to organize opposition, or they might come to monitor proposed activities and find themselves advocates. The lead agency or organizer also has interests, which can range from legitimizing a process by displaying community engagement to reducing dissent or empowering residents to achieve systemic change (White 2011). Therefore recognizing that everyone participates for both self- and collective interest can increase the commitment to working towards projects that work for everyone.

COMMON PARTICIPATORY TECHNIQUES

The scale of the project greatly influences how public participation processes will unfold. Larger projects that affect more people require broader participatory processes, whereas one small site or temporary activity might need only local outreach and discussion. For large projects over long time periods, various mechanism will ensure a wider range of residents receive information and can be involved. It is important that the mechanisms reflect the range of stakeholders, target those who are or will likely be underrepresented, and allocate the resources necessary to conduct the process well. It is equally important to choose the mechanisms that best accomplish the intended objectives (Table 14.1).

Public meetings, hearings, and workshops

Public meetings are a common form of participation in the US. They are a primary form of information-sharing where lead agencies or organizations provide information to participants and receive feedback. Participants' perspectives can be gathered through a public comment period, a questionnaire, or another tool that allows participants to convey what they think. Public workshops usually involve a formal discussion component, where participants talk for longer periods, often in small groups, in order to better develop and share ideas. A public hearing might be required, where a hearing body such as a planning commission listens to staff members' recommendations on a particular project, the developers' presentations, and community members' questions. This forms the basis of a decision by the commission. The advantage to public meetings is they have the capacity to include all interested people and they can accommodate all who want to participate. The limitation is that they rarely have formal mechanisms to synthesize or incorporate public comments into the final plans or projects, and there is no legal obligation to do so. In an analysis that looked at the participatory processes in 239 cases of environmental decision-making, 21% used some form of public hearing or meeting (Beierle and Cayford 2002: 44).

Advisory committees

Advisory committees are another common way to bring stakeholders together. Out of the 239 environmental decision-making cases mentioned above, 55% had an advisory committee (Beierle and Cayford 2002: 44). In this case, stakeholders are invited or nominated to participate, and they are assumed to represent a community or other interest. They can include local residents and community leaders as well as environmental and social organizations or industry. In the case of advisory committees, the same participants work together throughout the process in order to develop mutually beneficial outcomes and advise the lead agency or organization. They are granted authority by the lead agency, which might include making recommendations or making decisions, but their findings are not legally binding.

Advisory committees allow for more intensive engagement, where the advisory committee members become very knowledgeable and make recommendations that are both technically sound and incorporate public values. When stakeholders with conflicting positions work together, they can also reach mutually beneficial outcomes without litigation or formal negotiation. However, it is also possible that, to reach consensus, either important issues or stakeholders are left out of the discussion (Beierle and Cayford 2002).

Negotiation and mediation

Negotiations and mediations are entered into when stakeholders have differences that are difficult to reconcile. These often come about when litigation is instigated or threatened. Negotiations

TABLE 14.1 PARTICIPATORY TECHNIQUES AND ANTICIPATED OUTCOMES

Technique	Participant base		Type of engagement		Capacity to:											
					Incorporate local knowledge		Incorporate local priorities		Educate and inform		Build trust		Resolve conflicts		Monitor or stop destructive action	
	Narrow	Broad	Information sharing	Deliberation	High	Low	High	Low	High	Low	High	Low	High	Low	High	Low
Public meetings		x	x		Varies		Varies		x			x		x		x
Public hearings		x	x		Varies		Varies		x		Varies		Varies		Varies	
Public workshops		x		x	x		x		x		x			x		x
Advisory committees	x			x	x		x		x		x		Varies		Varies	
Negotiation and mediation	x			x	x		x			x		x	x		x	
Online tools		x	Varies		Varies		Varies		Varies		Varies		Varies		Varies	

involve representatives of stakeholders and organizations to develop an agreed-upon course of action. Negotiations and mediations reach legally binding agreements, unlike advisory committees or public meetings, but they usually address a small slice of the common issues and come about when issues cannot be resolved. Of the 239 environmental decision cases, 24% involved negotiation or mediation (Beierle and Cayford 2002: 46).

Online participation

Lead agencies and organizations have increasingly used online participatory techniques to augment conventional face-to-face public meetings. Online mechanisms can include information-sharing via email or a website. They can provide information affordably to people as well as collecting comments or survey responses about important issues. Online streaming of meetings with a mechanism for feedback also offers an additional way for people to be involved. In these cases, the intent of online participatory tools is not different from other mechanisms, but they can reduce constraints placed by time and distance that may prevent people from participating.

Community members, as well as lead agencies and organizations, have experimented with different platforms to better engage more people to participate in envisioning their future. They have created interactive platforms to allow people to visualize new proposals and give their ideas form in real time—and to allow people to vote on others' ideas, creating a form of crowdsourcing for urban solutions. The public space game *Crowdsourced Moscow 2012*, for example, gave participants opportunities to make proposals for public space use and design and to work with others to negotiate conflicts or meet collective goals (Desouza 2012). Simulated environments have potential to show people the outcomes of different options or proposals in more immersive ways than traditional design drawings. Online participatory tools nevertheless raise issues—the outreach must reach a broad base of participants, people need good information to participate and make recommendations, and lead agencies must have ways to incorporate or make decisions among different ideas. In addition, not everyone is equally comfortable using online tools, and it is not yet clear that the platforms will help with intensive discussions about conflicting goals (Desouza 2012). Although online platforms offer alternatives, they do not yet replace traditional participatory tools.

THE CHALLENGES AND PROMISE OF PUBLIC PARTICIPATION

In this section we discuss challenges that are inherent to participation in complex, inequitable societies. Although there are no easy solutions to these challenges, awareness of these issues makes participatory activity more likely to meet multiple objectives.

ACHIEVING DIVERSE PARTICIPATION AND REPRESENTATIVENESS

A primary challenge to participatory processes is engaging the diverse groups and varied interests of people who will be impacted. Participatory processes never perfectly reflect the community's diversity, and wealthier and more privileged residents often participate at higher rates. As a result, directed outreach to diverse community members is necessary to achieve diverse participation.

Outreach can take different forms. It might include holding meeting and information sessions in spaces where community members will be more likely to participate, such as at a community center, a workers center, or even a restaurant. Organizing meetings in partnership with local institutions, such as neighborhood organizations, churches, tribal governments, senior housing complexes, or childcare centers, can increase the number of people involved. Initial outreach to visible community leaders, including nonprofit organizations, local business owners, church pastors, and staff members of community-based organizations, can help jumpstart a dialogue before establishing a full participatory process.

Because it is unlikely that any participation process will proportionately reflect the community, using field methods to gather information before and after community engagement processes can increase the knowledge base when starting, and fill in gaps if particular populations do not participate. With diverse participation comes diverging views, and processes must be designed to incorporate and respect varied perspectives.

Lead agencies and organizations often seek to have representative participation or people who represent different stakeholder positions. In some cases, representation can mean being authorized to act or speak for someone, such as a member of an organization. It also might mean being accountable or answering for one's actions. This is the case with public officials who represent their constituents. Representation also can refer to standing in for, or reflecting, a particular position, which is what planners assume when they seek demographic diversity. It also can suggest acting in the interest of other similarly situated people, which planners assume when particular stakeholders are recruited to participate (Thomson 2009).

In practice, what constitutes representation has been debated for decades (Alexander 1976) because no one person shares all characteristics and perspectives with others. The lead agency or organizations must make intentional decisions regarding outreach while realizing that community members will have different perspectives about who can represent them.

POWER, INJUSTICE, AND SITUATED PERSPECTIVES

Some forms of diversity also reflect past injustices that continue to operate in contemporary social structures. As a result, ordinary urban processes often perpetuate inequality because deeper injustice can be embedded in what appears to be a discrete situation. Before collective action can be taken, this needs to be made visible and acknowledged by all participants (Sandercock and Lyssiotis 2003). It is impossible immediately to alleviate the deep injustice that resulted in different Native Americans losing their land throughout the US and Canada, or the structural inequalities that produce mass incarceration. However, each action can attempt to serve as a step towards better and more just future settlement. For example, white residents might wonder why a discussion about race continues, and organize a community meeting with the intent to "get over race," a meeting that might be manipulation. Instead, the energy can be focused on how the current situation perpetuates and reflects racial inequality, and what can be done to dismantle the structural inequalities.

Similarly, the power imbalance can exacerbate conflicting perspectives. In some circumstances, participants bring up issues that appear at first glance to extend outside the current discussion. Oftentimes, the professionals—whether they represent developers, the local business community, or the public sector—express impatience with community members' unwillingness to stay on topic. For instance, in the 2010s, a proposed jail expansion public discussion directed attention to injustice in the criminal justice system and the problems facing families with incarcerated members. Although this was "off topic" for the professionals, who separated the development project from the system of mass incarceration in the US, ordinary citizens were unwilling to separate the jail as a development project from the criminal justice system as an institution that impacted their daily lives. Activists also were unwilling to separate the number of beds planned from the processes leading to mass incarceration.

As this example illustrates, the concerns that community members raise may be local manifestations of deeper social and economic structural problems and global processes. This occurs repeatedly. When lower-income community members express skepticism that a new project will hire local residents, regardless of promises to do so, this perspective reflects their experiences with the larger labor market. Similarly, when middle- and upper-income community members state that they prefer independent businesses to chains, and fear that new developments will bring chain retailers that will displace local businesses, as has happened throughout the country, they are reflecting on a trend that extends beyond the proposed project.

Through field methods and engaging with regular people, revitalization professionals might come to realize the depth of divisions in an area, or the legacy of fighting for justice and local improvement that has led to mistrust of the public sector and developers. They might also realize that it is difficult to help a community gain new development which is consistent with its vision because of how market forces play out. Despite being empathetic, it might often appear impossible to fully address the concerns that a given project raises.

ANTICIPATING MISTRUST AND BEING ACCOUNTABLE

Because of the complexity of urban change, and because many people are mistrustful both of developers and of government, revitalization processes are riddled with distrust. In many cases this is entirely justified. The history of urban development in the US is filled with unjust decision-making, which has fostered significant mistrust in some communities. Anticipate this mistrust. Rather than understanding trust as an object to achieve, trust might be better thought of as a set of dynamic relationships that are strengthened and weakened by the level of transparency and accountability in the planning processes.

The "for what?" question becomes critical because participatory processes should be aligned with issues around which the participants can have influence. However, regular people are often at a socially constructed disadvantage. For example, if a developer is planning a project on a lot that community members want for a less profitable use, such as a park or a community garden, discussions about the park may create a situation where participants' ideas are discarded and participatory processes will likely have little influence. Developers and their representatives are steeped in property law and finance, and have considerably greater financial resources and business connections than most community groups. Before going public, developers do a great deal of research to assess a project's feasibility and often invest considerable funds and political capital to ensure an outcome in their favor. While developers should be held accountable and required to incorporate residents' needs into a project, more often the project scope is determined by legal procedures and return on investment. Unless there are legal stipulations for participation and community determination, communities of regular people are at a disadvantage.

Irazábal and Punja (2009) illustrate this well in their case study of the lawsuit surrounding the displacement of the South Central Farm. The Farm was a 14-acre community garden established in a Los Angeles neighborhood defined by a high level of industrial land use and concentrated poverty. As they show, our system of land-use law contains a built-in structural bias that favors the developers' claims, rooted in the exchange value of the land. After ignoring the property for years as the farmers established their community space, the developer eventually sought to profit from the increased property values on the site due to the growth of the adjacent Port of Los Angeles. The property owner had the legal right to evict the farmers despite the improvements and investments they had made to the land in his absence. In other words, the legal system simultaneously backed the exchange value of the property as it delegitimized the claims of the farmers' use of the land as a garden and green space in an impoverished industrial area.

Nonetheless, community organizations and coalitions have been able to negotiate with private and public sector coalitions in some circumstances. For example, local and national groups such as the Los Angeles Alliance for a New Economy (LAANE) and Good Jobs First promote and organize around the concept of "accountable development." Well-known examples of accountable development are the anti-Walmart and big box campaigns. Groups such as LAANE have been instrumental in bringing together diverse coalitions of people to speak out against the negative impacts of big box stores on jobs, wages, traffic, and small businesses. Their efforts have led to ordinances restricting the size and siting of big box stores and have put pressure on Walmart and other mega-retailers to improve their hiring practices and community outreach.

Many situations arise where particular interests will be privileged over other concerns. Property ownership is one particularly tricky issue in terms of representativeness and participation. For example, in some neighborhoods, many property owners do not live in the area. Traditionally, these have been absentee landlords who rent to others. In some cases, neighborhoods also might have part-time residents (who live most of the year elsewhere) or high numbers of vacation rentals. To what extent should they have a voice in the revitalization planning process? Or, do squatters in vacant property or people living under a freeway overpass have a say? Do longtime residents' perspectives carry more weight than newcomers'? In practice, property owners' interests are often privileged even at the expense of local residents. Figuring out how to be responsive and accountable while recognizing that you work in larger economic and social structures is a significant and important challenge to consider.

WORKING FROM THE GRASSROOTS AND THE OUTSIDE

In many situations, residents take the initiative to stimulate change—or engage in bottom-up revitalization efforts. But as often, residents find that official processes do not reflect their interests adequately, or they learn late about a proposal that will impact their neighborhood. The objectives of such community change efforts are both to impact the material conditions of neighborhoods—therefore they include revitalization agendas—and to build civic capacity to effect future change. While organizing efforts can culminate in revitalization initiatives, equally as often they intersect with specific revitalization projects initiated by an outside agency or organization. Residents might oppose a proposal in their neighborhood, even if it purportedly will stimulate economic development.

Community organizing seeks to mobilize a group around their shared interests and issues. Community organizers typically focus on specific issues, such as access to adequate education or job training, as a means to organizing a grassroots coalition. Community organizing can be a long-term process, but at times residents organize quickly in response to revitalization agendas that threaten their community. Community organizing has challenged revitalization initiatives through confrontational actions and campaigns to stop proposed projects, with mixed successes.

The most common technique is to use confrontational or disruptive tactics to draw attention to an injustice and put pressure on public officials or corporate entities to meet their demands (Alinsky 1969, Saegert 2006). For example, the Mission Community Organization in San Francisco infamously withdrew money from a bank using tortillas as checks to protest discriminatory hiring practices (Castells 1983). Once these efforts gain attention, organizers seek to engage their target in more productive negotiations. Despite the outward appearance of confrontation, however, internally these organizations rely on strong social connections and participation (Saegert 2006). The Texas Industrial Area Foundation, a highly successful community organizing network, practices a "relational organizing strategy which emphasizes teaching people how to connect to each other, to build a relationship that can empower them to act" (Warren 2001).

In Atlanta during the 1960s, a news item about displacing 966 families in the Bedford-Pine neighborhood, which had been designated an urban renewal era, sparked neighborhood activism led by three ministers and two merchants. They chose not to act through the city council but to engage residents directly through a formidable grassroots group, U-Rescue. U-Rescue made various demands, including that the area remain residential and that a citizens' advisory council be established. The city made concessions, requiring that public housing be built in the area, even though this was against the wishes of a major hospital, and required housing development would occur in stages to reduce development impacts. Although U-Rescue gained notable concessions, the long-term outcome was mixed—displacement still occurred, but the area retained its residential character (Stone 1989). In many cases, growth coalitions make headway despite community opposition, which has furthered the distrust discussed above.

Responding to well-organized communities without paying attention to power dynamics can also lead to inequitable outcomes. Middle- and upper-income communities have great capacity to organize quickly if they perceive a threat. This can result in unpopular facilities being located in areas where resistance will gain less momentum because the residents are poorly organized, have fewer resources, or receive less public support for their concerns. For example, in the late 1960s in New Orleans, a proposed waterfront freeway that would have cut through the historic French Quarter was successfully opposed by a coalition of predominantly white residents and historic preservation advocates. The freeway spur was then constructed through the historic Tremé neighborhood along a prominent African American commercial corridor. Revitalization professionals must be responsive to organized community groups while ensuring that in doing so projects do not disproportionately hurt residents with a weaker voice.

Although public processes intend to give ordinary citizens a voice in revitalization processes, it is important to realize that people continue to work from the outside as another way to shape their cities and neighborhoods in desirable ways. Because actions outside of and even against proposed revitalization projects make community concerns visible, they continue to be legitimate expressions of engagement that revitalization professionals confront.

SUMMARY

Public participation is a democratic ideal that proposes people should participate in decision-making when actions will impact them. Because development projects and other revitalization programs are intended to affect people, public participatory processes are an important component. From localized outreach when a temporary festival is planned to sustained engagement in major redevelopment projects, critical questions about participation include considering who participates, when they participate, how they participate, and what are the outcomes of the participation. People will become involved when it is in their interest to do so. However, the onus is on the organizers to ensure participatory processes engage the diverse residents and different stakeholders, and grapple with the questions that make participation as challenging as it is necessary.

STUDY QUESTIONS

1   Is Sherry Arnstein's ladder of participation still relevant? Why, or why not?
2   Attend two public meetings in your community. See if you can identify one that takes place within the neighborhood and one in a more official venue. Analyze the meetings and explain what worked well and what did not.
3   What types of public participation are required for development projects in your city? How are residents notified? By what mechanism? Will the requirements lead to sustained engagement? Why, or why not?

REFERENCES

Alexander, Chauncey A. 1976. "What does a representative represent?" *Social Work* 21(1): 5–9.
Alinsky, Saul. 1969. *Reveille for radicals.* Chicago, IL: University of Chicago Press.
Arnstein, Sherry R. 1969. "A ladder of citizen participation." *Journal of the American Institute of Planners* 35(4): 216–224.

Beierle, Thomas C and Jerry Cayford. 2002. *Democracy in practice: public participation in environmental decisions*. Washington, DC: Resources for the Future.

Castells, Manuel. 1983. *The city and the grassroots: a cross-cultural theory of urban social movements*. Berkeley, CA: University of California Press.

Cohen, John and Norman Uphoff. 2011. "Participation's place in rural development: seeking clarity through specificity." In *The participation reader*, edited by Andrea Cornwall. London: Zed Books.

Cooke, Bill and Uma Kothari. 2001. *Participation: the new tyranny?* London: Zed Books.

Cornwall, Andrea. 2008. "Unpacking 'participation': models, meanings and practices." *Community Development Journal* 43(3): 269–283.

Cornwall, Andrea and John Gaventa. 2001. *Bridging the gap: citizenship, participation and accountability*. PLA Notes 40. London: IIED.

Day, Diane. 1997. "Citizen participation in the planning process: an essentially contested concept?" *Journal of Planning Literature* 11(3): 421–434.

Desouza, Kevin C. 2012. "Leveraging the wisdom of crowds through participatory platforms." *Planetizen* March 5. www.planetizen.com/node/55051

Fainstein, Susan S and Scott Campbell. 2012. *Readings in planning theory*. Malden, MA: Wiley-Blackwell.

Fischer, Frank. 2000. *Citizens, experts, and the environment: the politics of local knowledge*. Durham, NC: Duke University Press.

Gans, Herbert J. 1982. *The urban villagers: group and class in the life of Italian-Americans*, updated and expanded edition. New York: Free Press.

Hanson, Royce. 2003. *Civic culture and urban change: governing Dallas*. Detroit, MI: Wayne State University Press.

Healey, Patsy. 2006. *Collaborative planning: shaping places in fragmented societies*. Basingstoke, UK: Palgrave Macmillan.

Irazábal, Clara and Anita Punja. 2009. "Cultivating just planning and legal institutions: a critical assessment of the South Central Farm struggle in Los Angeles." *Journal of Urban Affairs* 31(1): 1–23.

Marris, Peter. 1986. *Loss and change*, revised edition. London: Routledge.

Mohl, Raymond A. 2004. "Stop the road freeway revolts in American cities." *Journal of Urban History* 30(5): 674–706.

Rubin, Lillian B. 1969. "Maximum feasible participation: the origins, implications, and present status." *Annals of the American Academy of Political and Social Science* 385(1): 14–29.

Saegert, Susan. 2006. "Building civic capacity in urban neighborhoods: an empirically grounded anatomy." *Journal of Urban Affairs* 28(3): 275–294.

Sandercock, Leonie and Peter Lyssiotis. 2003. *Cosmopolis II: mongrel cities of the twenty-first century*. London: Continuum.

Saxena, NC. 2011. "What is meant by people's participation?" In *The participation reader*, edited by Andrea Cornwall. London: Zed Books.

Scott, James C. 1998. *Seeing like a State: how certain schemes to improve the human condition have failed*. New Haven, CT: Yale University Press.

Stone, Clarence N. 1987. "The study of the politics of urban development." In *The politics of urban development*, edited by Clarence N Stone and Heywood T Sanders, 3–22. Lawrence, KS: University Press of Kansas.

Stone, Clarence N. 1989. *Regime politics: governing Atlanta, 1946–1988*. Lawrence, KS: University Press of Kansas.

Thomson, Ken. 2009. *From neighborhood to nation: the democratic foundations of civil society*. Hanover, NH: Tufts University Press.

Vale, Lawrence J. 2013. *Purging the poorest: public housing and the design politics of twice-cleared communities*. Chicago, IL: University of Chicago Press.

Warren, Mark R. 2001. *Dry bones rattling: community building to revitalize American democracy*. Princeton, NJ: Princeton University Press.

White, Sarah. 2011. "Depoliticizing development: the uses and abuses of participation." In *The participation reader*, edited by Andrea Cornwall. London: Zed Books.

SECTION V

Looking Forward

15

Localism, regionalism, global governance, and beyond

WHAT WE KNOW: UNEVEN DEVELOPMENT AND URBAN REVITALIZATION POLICY

Uneven development is a defining feature of our cities. Throughout the dramatic changes that have occurred since the rise of the industrial city, the overarching purpose of urban revitalization policy has been to combat the negative effects of the always shifting landscape of uneven development. Historically, economic fortunes changed and private investment moved from one neighborhood or region to another. The result has been to enhance the prospects of some and hurt those of others. When governments channel resources to specific areas, they can affect positive change, but they can also contribute to the process of uneven development. Many times, their decisions help those in need of assistance, but frequently the result has been to improve the conditions in specific places and set into motion a process of displacement or exclusion. Various actors, from private developers and those in public office to everyday residents, make decisions that cumulatively open up or limit opportunities in different places, from neighborhood blocks to regions and nations. This process of "creative destruction" reconfigures places, their built environments, businesses, and life opportunities.

This is where urban revitalization policy enters the development process: to mitigate against the negative effects of uneven development and work to ensure that all can flourish. In practice, urban revitalization policy has often been focused on creating market opportunity and adjusting to landscapes of uneven development without working directly to create opportunity and reduce structural inequality.

As history shows, there is no set formula for urban revitalization. Over the nearly 70 years since the initiation of the federal urban renewal program, we have witnessed a shift from large-scale demolition and rebuilding efforts in the urban core to more comprehensive community initiatives that attempt to simultaneously address housing, economic development, health, and education to improve life in specific places. We have seen a variety of programs emerge to recover rather than destroy industrial buildings and infrastructure. Some are very visible private sector initiatives to redevelop highly valuable tracts of land in the central cities. We have also seen the rise of smaller, more incremental community efforts around housing, social engagement, and temporary uses of overlooked public space. While all of these have been initiated in the name of urban revitalization, outcomes have varied over time, place, and approach.

Urban revitalization policy trends change over time, and many revitalization strategies are replicated from place to place based on their perceived success. On the surface, this makes sense since downtowns, struggling inner-suburban and central city areas, shrinking cities, and gentrifying neighborhoods all wrestle with the effects of uneven development. However, each of these places faces varied revitalization challenges, and they do so in different ways and with different assets and actors. For instance, shrinking cities grapple with challenges such as abandoned, vacant and brownfield properties on a large scale while other cities may battle with the effects of too much development. These cities must work to ensure that neighborhoods with good services and accessibility are affordable to a range of incomes. Places also have different local histories and varied networks of economic and cultural exchange. Some suburban areas have watched as businesses and population move to both the outer suburbs and central cities and face the challenge of attracting investment to retrofit empty big box stores and dead malls. Others have become home to flourishing immigrant communities that initiate the process of revitalization and retrofit. Yet these communities may face challenges in adapting to a new life and potential discrimination from longtime residents. For these reasons, urban revitalization efforts must respond to broader economic forces and social trends as well as the issues in a particular locality.

THE GOAL: TO CREATE JUST AND LONG-LASTING CHANGE

Despite these different contexts and challenges, the overarching goal of urban revitalization remains the same: to create just and long-lasting change that benefits people and their communities. Within the iterative exchange between global forces and local action, the ongoing challenge with revitalization efforts is to structure each project so that neighborhoods and their residents will not become disadvantaged with each turn of the economy. Instead, we seek revitalization initiatives that can dismantle historic patterns of disadvantage and replace them with fair and plentiful opportunities that lead to prosperous and healthy places. As we laid out in chapter 1, urban revitalization projects can address multiple urban challenges. They can contribute to building human capital and enhancing social–cultural equity. They can improve place attractiveness and economic competitiveness. They can also create vibrant built environments and contribute to more environmentally sustainable places. Because rapid economic shifts leave behind specific regions, resulting in significant economic hardship and environmental damage, the objective is to respond to systemic change while addressing specific, local concerns.

Therefore another major challenge of urban revitalization policy is to integrate the long-standing emphasis on place-based initiatives with those that provide benefits and assistance to people directly, regardless of where they live. Urban revitalization not only should be concerned with adapting the built environment to be more accessible, equitable, and valuable, but also should consider how to improve the education, employment, and life skills of individuals. Place-based revitalization will remain crucial because where we live and work has a tremendous impact on our outlook and opportunities. We also need to enhance place-based efforts by connecting people to opportunities outside their neighborhoods and enabling them to feel empowered to make change.

TAKING STEPS TO BUILD SYSTEMIC CHANGE

How can local actors who are working to revitalize their neighborhoods ensure that the benefits are broadly realized? As the previous chapters have shown, trends that led to disinvestment and reinvestment are shaped by intentional responses by urban residents to their existing conditions as well as their visions for the future. They respond to larger processes beyond their immediate control but also have a hand in shaping local outcomes. A key aspect of urban revitalization policy will be to enhance efforts to build civic capital, or the knowledge and skills about how to shape urban processes, engagement, and outcomes. Additionally, because urban revitalization efforts will continue to concentrate on particular places, we need to focus on expanding and reforming property markets, but we also need to consider ways in which people affect change at different scales.

BUILDING CIVIC CAPITAL

Across the US, it is not only government and private developers who have a role in shaping our cities. Regular people are also at the forefront of creating projects that positively transform their neighborhoods. Many of the examples we have discussed, from the initiation of temporary parks and urban farms to spearheading major efforts to remediate brownfield sites and reuse industrial infrastructure like the High Line, were due to the efforts of regular people. They have made political demands to participate in planning for large-scale redevelopment projects and have a say in important land-use decisions. In so doing, they draw on civic capital. Residents and other regular people use their skills and know-how to develop new ideas and projects that work locally and are transferable to other places. Urban revitalization efforts can bolster technical assistance in ways that give more people the ability to make change and incorporate engagement in ways that help to facilitate and grow civic capital. As we discussed in chapter 14, engaging residents and other community-based stakeholders is critical to achieve better project outcomes.

EXPANDING AND REFORMING PROPERTY MARKETS

A recurrent issue is that the built environment changes more slowly than economic systems and social trends. This repeatedly causes problems with vacant land for residents who continue to live in areas that lose economic activity and population. Acknowledging that land and property function differently from other tradable goods is a first step toward developing tools and policies that help residents in depopulating neighborhoods, cities or suburbs take control of their communities. As we discuss in chapter 1, private property has both use value and exchange value. Urban revitalization strategies can do more to address these often competing claims on property and place. Shrinking cities are at the forefront of rethinking these arrangements in order to make it easier for residents and business owners operating in the area to shape their surroundings. Programs that allow residents to acquire and maintain abandoned property are but one example. In stronger housing markets, community land trusts and other forms of collective ownership can create affordable housing options. Inclusionary housing and fair share housing programs are other initiatives to intervene in property markets to preserve use values. Still, we need to develop other innovative approaches to address this inherent challenge.

LOCALISM AS A RESPONSE TO UNEVEN DEVELOPMENT

Localism has been a prominent response to changes and constraints that have arisen from living in a globally integrated economy. The global flow of goods, services and people is a dominant characteristic of the contemporary era. The global economy provides a range of benefits that affect our everyday lives, but it also has led to homogenous products and corporate-dominated economies where profits from local sales primarily benefit people outside the region. As a result, local producers have a hard time selling their products to local markets and many are forced to shut down.

In response, many people have turned to buying locally produced goods and services precisely because they value the quality and uniqueness lost in the system of global commerce. This has focused attention, for example, on local food systems. Increasing numbers of people in the US substitute produce, meat, and cheese that are shipped globally with those that are produced locally, for example, within a 200-mile radius. This can create new local jobs in food production and, by reducing the distances that goods are shipped, curtails pollution. No regions will be entirely dependent on local food, and access to global products greatly increases food choices and reliability. Nevertheless, local food can provide seasonal and specialized food and augment what is provided through the global food production and supply chains (Hess 2009).

Another aspect of localism draws on the specialized crafts and skills of different regions. Here, localism refers to the unique place where the products are made. The growth in craft beers is an example. In 2013 craft beers were a $14.8 billion industry in the US, up from $11.9 billion in 2012 (Brewers Association n.d.). Many regions have developed craft breweries with strong reputations that enable them to sell to larger markets. Some areas have traditions in growing and producing food or other products, while others have more recently established niche markets in textiles, furniture or ceramics. Because the markets are growing for products that are handcrafted or made in small batches, this creates an opportunity for specialized production or locally made products to be distributed over large, even global, markets. These locally rooted markets can provide stable, high-quality employment that is not easily outsourced. Unlike traditional manufacturing industries that compete on cost or volume, local craft and small-scale manufacturing compete on innovative design and specialize in customized production that is associated with a specific place or region. As a result, these firms tend to be highly placebound and locally integrated.

Localism also recognizes and takes advantage of the potential for substituting a portion of local consumption in products made by global producers with local sources. Supporting local businesses increases the percentage of consumer spending dollars that stay in the region. In conjunction, independent businesses keep a higher percentage of expenditure circulating regionally and purchase supplies locally. Local energy production can also address environmental concerns while building a local economic base.

Nevertheless, urban revitalization organizers and professionals must approach local markets and production with an awareness of their potential pitfalls. Without specific attention to ensuring that local production also provides living-wage jobs and safe work environments, the vulnerabilities that people face when working in global food and craft production can be replicated at the local level. Global systems have created inexpensive food and products that make it more challenging to create living-wage industries while producing competitive products. It is critical to focus on high-quality jobs. In addition, many argue that purely consumerist solutions will not eliminate the externalities that markets generate, and care must be taken to ensure that local manufacturing industries become greener and more innovative.

REGIONALISM AS A LOCALIST RESPONSE

Localist responses also require regionalist strategies. Given the structure of the metropolitan region, in many ways localism is a form of regionalism, because many solutions become available at scales larger than the neighborhood or city. However, regionalism also refers to coordination among municipalities in shared governmental decision-making. For example, this may occur when designing incentive programs or transportation and environmental plans. Developing regional strategies has been hard to accomplish, particularly in areas where some cities are faring worse than others. Yet regional coordination among municipalities is also necessary to reduce the race-to-the-bottom effect that occurs when local governments compete with incentives to attract businesses or to develop effective green infrastructure systems. Because regionalism is critical, observers have not given up on regional strategies to counter the most devastating impacts of the fickle global economy and stabilize areas that are not well situated in a global marketplace (Imbroscio 2011).

Most regions have metropolitan planning organizations that offer a starting point to strengthen this planning scale. In addition, many state governments, including Washington and Florida, have required regional thinking in their growth management legislation through inclusive zoning requirements to help ensure all municipalities provide affordable housing, and growth boundaries that restrict outward growth while encouraging infill. California SB375, the "anti-sprawl bill" discussed in chapter 8, is another example.

Another reason to pay attention to regional strategies is that local revitalization policies can have regional effects. Urban renewal and other place-based revitalization initiatives such as HOPE VI have displaced residents. In some cases, dispersing residents has been one of the stated objectives under the rubric of deconcentrating poverty. In direct response, many have questioned whether redeveloping areas for new rather than existing residents is just or effective as a policy initiative (Fainstein 2010, Imbroscio 2011). Instead, an alternative framework is to use place-based initiatives to ensure everyone in that place can move up instead of out. This challenges revitalization organizers and professionals to build in affordability to all areas—for instance through incentive zoning, discussed in chapter 7—and create pathways to opportunities for residents residing in areas with high levels of disinvestment and underemployment, as we mention throughout this book. Building regional systems that rely more heavily on local businesses alongside building capacity in local residents can result in more dollars circulating locally, strengthening an area's capacity and regional economy.

THE CONTINUING ROLE OF THE NATION STATE AND THE POTENTIAL FOR GLOBAL GOVERNANCE

As we have discussed throughout this text, the contemporary period is defined by neoliberal state policies that look to market mechanisms and market efficiency to effect change. The transition to a globally integrated economy in a neoliberal political era resulted in federal policies becoming more market driven and less focused on redistribution to further social welfare goals. At the same time, federal government has taken a less prominent role in urban policy (Brenner 2004).

However, this does not mean that the nation state is unimportant. Even in the era of neoliberal globalization, cities and regions with weak markets that are located in countries with stronger urban policies have been better able to negotiate with the private sector to bring in economic development that serves residents well. Residents in countries with stronger urban and welfare policies also have not been as disadvantaged when local economies change. US urban and welfare policies are weaker than those of many of its peer countries and, as a result, weak market cities and many of the people who live there have struggled (Savitch and Kantor 2002). National policy, however, is dynamic and can change, as we have seen through the past 70 years of federal urban engagement. Although the national government has changed under neoliberalism, it is still a powerful institution that influences social welfare. It is imperative that people with urban revitalization expertise take advantage of national policies as well as advocate for policies that better help them achieve their goals.

In a global world, many regional actors are connected to global institutions and corporations. In order to reverse a global race to the bottom—as corporations move seeking lower wages and fewer environmental restrictions—political links through consumer pressure on corporations or political pressure through global governance institutions comes through civil society, where people in one locale are linked with others in similar situations throughout the world (Castells 2008). Attracting residents from other regions or parts of the world and creating open-minded societies is one step. Many revitalization efforts seek to attract global firms or retail such as Walmart. Walmart has been an inconsistent revitalization partner. It can self-finance its structures, making it a desirable business, but it also leaves empty big box stores when a new site becomes more profitable. Organizing across boundaries in a global corporate world is an effort to help make corporations act responsibly in our communities as well as in others.

A STARTING POINT FOR REVITALIZATION PROFESSIONALS AND ORGANIZERS

Revitalization organizers and professionals work at the intersection of local change at a project level and systemic change for the community, city and region. Whether the area is growing or shrinking, revitalization initiatives can help neighborhoods thrive and aid residents with different resources, opportunities and perspectives to find places for themselves. In order to accomplish this, urban revitalizers must pay close attention to linkages from a given locale to the regional system and beyond.

When approaching a given project, certain questions can make these linkages more visible. The first questions address the place's situation. How is the neighborhood situated in the city? How is the city situated in the region? How is the region situated in the global economy?

The second set of questions addresses who the project will impact. What are the characteristics of the people who will be impacted? Who is most likely to benefit? Who is most vulnerable? In what ways are they vulnerable?

The latter set of questions links the project to the area's situation and the people within it. What will the project accomplish? What are the mechanisms within it to create specific opportunities (e.g. housing or jobs) for residents who have fewer resources? What are other opportunities to link the project with broader objectives such as green infrastructure or regional urban growth strategies? What mechanisms help ensure stability in times of market shifts or when global actors find better opportunities elsewhere? By being mindful of the area, the people, and the project's broad potential, each project can be a step towards creating just, sustainable and prosperous communities.

REFERENCES

Brenner, Neil. 2004. *New state spaces: urban governance and the rescaling of statehood*. Oxford: Oxford University Press.

Brewers Association. 2014. "Brewers Association announces 2013 craft brewer growth." Press release March 13. www.brewersassociation.org/press-releases/brewers-association-announces-2013-craft-brewer-growth

Castells, Manuel. 2008. "The new public sphere: global civil society, communication networks, and global governance." *Annals of the American Academy of Political and Social Science* 616(1): 78–93.

Fainstein, Susan S. 2010. *The just city*. Ithaca, NY: Cornell University Press.

Hess, David J. 2009. *Localist movements in a global economy: sustainability, justice, and urban development in the United States*. Cambridge, MA: MIT Press.

Imbroscio, David. 2011. *Urban America reconsidered: alternatives for governance and policy*. Ithaca, NY: Cornell University Press.

Savitch, Harold V and Paul Kantor. 2002. *Cities in the international marketplace: the political economy of urban development in North America and Western Europe*. Princeton, NJ: Princeton University Press.

Index